Structural Textile Design

Structural Textile Design

Interlacing and Interlooping

Yasir Nawab, Syed Talha Ali Hamdani,
and Khubab Shaker

CRC Press
Taylor & Francis Group
Boca Raton London New York

CRC Press is an imprint of the
Taylor & Francis Group, an **informa** business

CRC Press
Taylor & Francis Group
6000 Broken Sound Parkway NW, Suite 300
Boca Raton, FL 33487-2742

First issued in paperback 2020

ISBN 13: 978-0-367-57372-0 (pbk)
ISBN 13: 978-1-4987-7943-2 (hbk)

Library of Congress Cataloging-in-Publication Data

Names: Nawab, Yasir, editor. | Hamdani, Talha, editor. | Shaker, Khubab, editor.
Title: Structural textile design : interlacing and interlooping / edited by Yasir Nawab, Talha Hamdani, and Khubab Shaker.
Description: Boca Raton, FL : CRC Press, 2017. | Includes bibliographical references and index.
Identifiers: LCCN 2016043396| ISBN 9781498779432 (hardback : alk. paper) | ISBN 9781315390406 (ebook)
Subjects: LCSH: Textile fabrics.
Classification: LCC TS1765 .S77 2017 | DDC 677--dc23
LC record available at https://lccn.loc.gov/2016043396

Visit the Taylor & Francis Web site at
http://www.taylorandfrancis.com

and the CRC Press Web site at
http://www.crcpress.com

Contents

Section I   Woven Fabric Structures

Section II   Knitted Fabric Structures

Preface

Fabric is one of the basic requirements of humans not only for clothing but also for esthetic and functional purposes. There exist numerous fabric formation techniques, the most common being weaving and knitting. These techniques produce fabric by interlacing and interlooping, respectively. The fabric-forming process or fabrication method contributes to fabric appearance, texture, suitability for end use, etc. These fabrics have varying structures, depending on the interlacement/interlooping pattern of the yarns, which governs the properties of fabric. Varying the structural design of fabric has significant impact on the esthetics. Fashion industry also focuses on the design of base fabric to achieve the desired effects.

There is a vast area of application for fabric, as we look around; ranging from shirting to home textiles and technical textiles. The fabric formation techniques (weaving and knitting) result in fabrics that greatly differ in terms of structure. The yarns in weaving are relatively straight with some crimp, while the yarns in knitted fabrics form a looped structure. Therefore, the woven fabrics have less extensibility as compared to the knitted fabrics. A slight variation in the interlacement/interlooping pattern may result in a fabric with entirely different properties. Therefore, design of a fabric, whether woven or knitted, is an important factor in determining the physical and mechanical properties of the fabric.

This book aims to cover most of the designs used in the production of woven and knitted fabrics. This book has been organized into two major sections along with a preliminary portion consisting of two chapters. Chapter 1 of preliminary portion focuses on the introduction to textile fabrics, classification, application areas, and global trade. An insight into the CAD software used in the textile industry for woven and knitted fabrics has been provided in Chapter 2.

Section I focuses primarily on woven fabric structures, starting with an introduction to the basics of weaving technique. The conventional weave structures have been elaborated in Chapter 4, focusing on the techniques to draw and execute designs on the weaving machine. The fabric structure, its properties, and the ultimate area of application have also been discussed. Chapter 5 deals with specialty fabric structures like multilayered fabrics, shaped structures, and piled fabrics. Chapter 6 includes textile structures for jacquard weaving, while Chapter 7 acquaints the readers with the use of color and its application to produce an esthetically appealing fabric according to the requirement.

The most important feature of this book is the addition of warp and weft knitting designs that have been included in Section II. The published literature on the said topics is very limited. Section II starts with an introduction

to knitting (Chapter 8) followed by Chapters 9 and 10 dedicated to patterning in weft knitting and warp knitting, respectively. Chapter 11 highlights the color and stitch effect in the knitted fabrics.

Yasir Nawab
Syed Talha Ali Hamdani
Khubab Shaker

Acknowledgments

Praises to the Almighty who enabled us to transfer our knowledge for the education and wellbeing of mankind.

The editors would like to express their deepest gratitude to the chapter authors for the precious time they put in to complete this book. We are also obliged to all those who provided support for the improvement of the manuscript. Special thanks to Muhammad Zohaib Fazal, who encouraged and motivated the authors to meet deadlines. We are also thankful to Taylor and Francis Group for providing us the opportunity to write this book.

Editors

Yasir Nawab is an assistant professor at the National Textile University, Faisalabad, Pakistan. His research areas are polymers, composite materials, 2D and 3D woven and knitted fabrics, finite element analysis, and analytical modeling.

Syed Talha Ali Hamdani is an assistant professor at the National Textile University, Faisalabad, Pakistan. Recently, he won the HEC start-up grant for the development of nonwoven sensor.

Khubab Shaker is a lecturer of weaving at the National Textile University, Faisalabad, Pakistan. He completed his master's degree in textile engineering in 2014 and is currently enrolled as a PhD scholar. His research areas include 2D and 3D woven fabrics, computer applications in weaving and textile composite materials.

Contributors

Hafiz Shehbaz Ahmad
Department of Knitting
National Textile University
Faisalabad, Pakistan

Waqas Ashraf
Department of Knitting
National Textile University
Pakistan

Habib Awais
Department of Knitting
National Textile University
Faisalabad, Pakistan

Danish Mahmood Baitab
Department of Weaving
National Textile University
Faisalabad, Pakistan

Muhammad Zohaib Fazal
Department of Weaving
National Textile University
Faisalabad, Pakistan

Syed Talha Ali Hamdani
Department of Weaving
National Textile University
Faisalabad, Pakistan

Muzammal Hussain
Department of Knitting
National Textile University
Faisalabad, Pakistan

Madeha Jabbar
Department of Garments
    Manufacturing
National Textile University
Faisalabad, Pakistan

Haritham Khan
Department of Knitting
National Textile University
Faisalabad, Pakistan

Muhammad Imran Khan
Department of Weaving
National Textile University
Faisalabad, Pakistan

Adeela Nasreen
Department of Weaving
National Textile University
Faisalabad, Pakistan

Yasir Nawab
Department of Weaving
National Textile University
Faisalabad, Pakistan

Muhammad Umar Nazir
Department of Weaving
National Textile University
Faisalabad, Pakistan

Khubab Shaker
Department of Weaving
National Textile University
Faisalabad, Pakistan

Muhammad Umair
Department of Weaving
National Textile University
Faisalabad, Pakistan

1

Textile Fabrics

Madeha Jabbar

CONTENTS

1.1 What Is a Textile Fabric?

The term "textile" originated from the Latin word textilis and the French word texere, which means "to weave," and the term was initially used only for woven fabrics. The advent of new developments has also broadened the scope of this term, covering filament, fiber, and yarn which are capable of being converted into fabric, and the subsequent material itself. Textile, therefore, includes threads, cords, ropes, braids, lace, embroidery, nets, and fabrics produced by weaving, knitting, bonding, felting, or tufting. To be useful in textiles, fibers must have some desirable properties such as strength, abrasion resistance, flexibility, and moisture absorption.

Studies on the history of fabric production are based on archeological findings, pictorial representations, frescoes, stone monuments, archival documents, etc. Perhaps the earliest fabrics were produced by felting, that is, condensing and pressing the fibers (mostly cotton or wool) together in the form of sheet. The cultivation of flax is documented as early as 6000 BC (Kvavadze et al. 2009), and the evidence exists for production of linen cloth in Egypt around 5500 BC. The Egyptians used linen fabrics, woven in narrow width, in the burial custom of mummification. Evidence report silk production in China between 3000 BC and 5000 BC. Scraps of silk were found in China dating back to 2700 BC. Hundreds of years before the Christian era,

cotton textiles were woven in India with matchless skill, and their use spread to the Mediterranean countries. In the first century, the Arab traders brought fine muslin and calico fabrics to Italy and Spain. The Moors introduced the cultivation of cotton into Spain in the ninth century. Fustians and dimities were woven there and in the fourteenth century in Venice and Milan, at first with a linen warp. Little cotton cloth was imported to England before the fifteenth century, although small amounts were obtained chiefly for candle-wicks (Balter 2009). By the seventeenth century, the East India Company was bringing rare fabrics from India.

1.2 Classification

There exist numerous techniques for fabric formation, and these techniques are the most usual parameter to classify the textile fabric into woven, knit-ted, nonwoven, and braided, as shown in Figure 1.1. The conventional fabrics (woven, knitted, and braided) are produced in such a way that the fibers are first converted into yarn and subsequently into fabric. However, fabrics are also produced directly from the fibers, without being converted into yarn. Such fabrics are termed as nonwoven fabrics.

Woven fabrics are produced by interlacement of two sets of yarns perpen-dicular to each other, that is, warp and weft forming a stable structure, while knitted fabrics are made up of interconnected loops of yarn. The bent yarn in a loop provides stretch, comfort, and shape retention properties to knit-ted fabric. However, the knitted fabrics are generally less durable than the woven fabric. Such properties help to determine the end use of a specific fabric. The chemical and/or mechanical bonding or interlocking of fibers produces a fabric structure known as the nonwoven fabric. The process of fabrics formation also determines the name of fabric produced, for example, felt, lace, double-knit, and tricot.

The selection of method for fabric formation is largely a function of the properties desired in the fabric. In contrast, the cost of fabrication process is

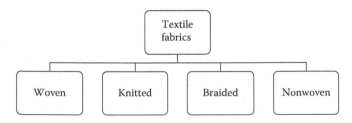

FIGURE 1.1
Classification of textile fabrics with reference to fabric formation techniques.

determined by the number of steps involved and the speed of production. The fewer the steps and the faster the process, the cheaper the fabric. Recent automations in the fabric formation techniques have resulted in improved quality, improved response to consumer demand, and made production more flexible. A more detailed overview of the two fabric formation techniques, weaving and knitting, is given in the subsequent chapters.

Another way to classify the textile fabrics is based on the raw materials used, as shown in Figure 1.2. The natural fibers are provided by Nature in ready-made form and need only to be extracted, whereas man-made fibers are generated by humans from the things that do not exist in the fibrous form.

Natural fibers may be defined as bio-based fibers or fibers from plant and animal origins (Nawab 2016). This definition includes all natural cellulosic fibers (cotton, jute, flax, hemp, sisal, coir, ramie, etc.) and protein-based fibers such as wool and silk. The mineral fibers (asbestos) are excluded by this definition, which occur naturally but are not bio-based. They are known to cause health risk and are prohibited in many countries. The man-made fibers are either cellulosic (viscose rayon and cellulose acetate), semi-synthetic (soy protein, bio-polyester, polyhydroxy alkanoate [PHA], poly lactic acid [PLA], and chitosan), or synthetic (polyester, nylon, aramid, elastane, polyolefins, etc.).

The natural fibers have been in considerable demand in recent years because they are eco-friendly and renewable. In addition, the natural fibers have low density, better thermal properties, and are biodegradable (Shaker et al. 2016). The man-made fibers on the other hand offer the advantage of tailored properties (physical or mechanical) and are also durable. A major problem with man-made fibers is their nonbiodegradability. The extraction of crude oil to manufacture man-made fibers creates toxic by-products that cause damage to the local environment. Currently, the researchers are working to address this issue, and they offer a green solution to these problems.

A large variety of textile products is being produced depending on the raw material, fabrication method, and the processes involved. These fabrics are used for a diverse range of applications across the globe.

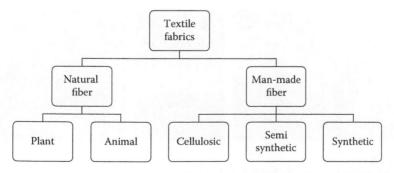

FIGURE 1.2
Classification of textile fabrics with reference to raw material.

1.3 Fabric Structure and Design

The fabric-forming process or fabrication method contributes to fabric appearance, texture, and its suitability for end use. The structures of woven, knitted, and nonwoven fabrics are shown in Figure 1.3. These fabrics have varying structures depending on the interlacement/interlocking pattern of the yarns. This sequence of interlacement/interlocking is termed as design of the fabric. The properties of fabric are largely governed by its design as well as the fiber content used as the raw material (Wulfhorst et al. 2006).

The woven fabrics are produced by interlacing two orthogonal sets of yarns, that is, warp yarns that are longitudinally arranged and weft yarns that are crosswise placed. The warp yarn is raised or lowered alternatively in a specific pattern over the weft yarn. This specific pattern for the distribution of interlacement is termed as the weave design of the fabric. While using CAD, the weave design is represented as an orthogonal array of binary numbers. For example, if warp yarn is over the weft yarn at crossover area, it is denoted by "1" and by "0" for the opposed case. In this way, an infinite number of weaves can be formed.

There are three fundamental weaves known as the basic weaves, namely plain, twill, and satin/sateen weaves, as shown in Figure 1.4. These basic weaves are characterized by small repeat size, ease of formation, and recognition (Adanur 2002). Plain weave is the simplest and has the smallest repeat size possible, that is, 2 warp yarns and 2 weft yarns. Derivatives of the plain weave are largely extension of weave in warp, weft, or both directions. Similarly, there exist a number of derivatives for the other two weaves, that is, twill and satin/sateen. The firmness of a woven fabric depends on the frequency of interlacement between warp and weft yarns. More the number of intersections per repeat, better the firmness of the fabric, while other

(a)   (b)   (c)

FIGURE 1.3
Fabric structure: (a) woven, (b) knitted, and (c) nonwoven fabric.

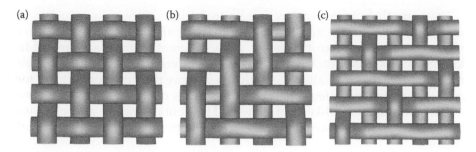

FIGURE 1.4
(a) Plain, (b) twill, and (c) satin fabric structures.

parameters are kept same. If we consider a plain and a twill woven fabric with same thread count and linear density, the latter will be less firmly compacted due to lesser bent than the former.

In addition to these basic designs, there are complex/intricate structures produced by the combination of these basic weaves, for example, multilayer fabrics, pile weave structures, and jacquard designs. These structures are widely used for a number of applications. The structure of multilayer fabrics is based on the stitching pattern of the individual layers. It is either layer to layer or through the thickness. The pile weave structures include the terry towels and woven carpets. In addition to the ground yarns, these structures contain an additional set of yarns called pile yarns, which appears on either one or both sides of fabric.

The knitted fabrics are produced by interconnected loops of yarn. Therefore, unlike woven fabric, the threads are not straight but follow a meandering path. These loops are easily stretched in different directions when subjected to external force, offering more elasticity than woven fabrics. Depending on knitting structure and yarn properties, the knitted fabric can stretch as much as 500% of its original length. Owing to this fact, the knitted fabrics are more form-fitting than woven fabrics and believed to be better suitable for garments that must be elastic or stretchable in response to the wearer's motions, such as socks and hosiery. The elasticity in knitted garments allows them to contour the body's outline more closely. The yarns used for knitting have normally low twist level, which gives them more bulk and less drape than a woven fabric.

The knitted fabrics are classified into warp-knitted and weft-knitted fabrics as shown in Figure 1.5. The yarns forming knitted fabric travel in a vertical direction in warp knitting (like warp threads in a woven fabric). In weft knitting, the yarn travels in a horizontal direction across the fabric. The weft-knitted fabrics can be produced either by hand knitting or by machine, whereas the warp-knitted fabrics are produced only on machines. The structural properties of the fabrics produced by these two different techniques vary considerably.

There are three basic stitches in a knitted fabric, namely knit, tuck, and miss. Different stitches and stitch combinations affect the appearance and physical properties of knitted fabric. Also, different patterns are created in knitted fabrics using stitches and yarns with different colors. For example, a row of tall stitches may alternate with one or more rows of short stitches for an interesting visual effect. Short and tall stitches may also alternate within a row, forming a fish-like oval pattern. Horizontal striping and checkerboard patterns (basket weave) may be produced by alternating knit and purl stitches.

Braided structures are constructed by interlacing one, two, or more sets of yarns with each other or with other sets of yarns at a certain angle. Half of this angle is usually equal to the angle between the yarns and the product axis (termed as the braiding angle). The patterns used in the braided samples are very limited, with fewer than five commonly used patterns. The braids may be produced either in tubular or in flat structure.

Nonwoven are the textile fabric structures produced from fibrous webs. The web may be bonded together by a number of means such as mechanical entanglement of fibers, use of adhesive, thermal fusion, and formation of chemical structures. The fundamental unit in the structure of a nonwoven fabric is the fiber, and the key factor is distance between the fibers. In general, the distance between adjacent fibers is several times greater than diameter of the individual fibers making up the web. Due to this reason, the nonwoven structures are more flexible. The characteristics of nonwoven fabrics can be manipulated easily and depend on the properties of the fibers used, the arrangement of the fibers in the web, and the properties of any binders or binding processes used.

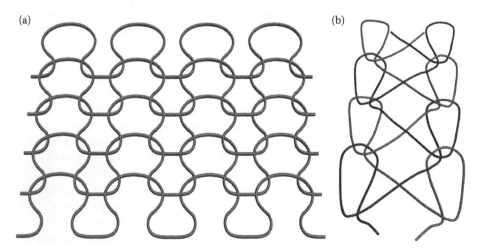

FIGURE 1.5
(a) Warp- and (b) weft-knitted structures.

1.4 Application Areas

As discussed in the earlier sections, the structure of textile fabrics varies greatly depending on the formation technique and pattern. The fabric structure determines the properties of these fabrics. On the basis of fabric properties, three major areas of application can be recognized, namely apparel, home textiles, and technical textiles.

1.4.1 Apparels

The term apparels is used for any type of fabric that is used for clothing. Application areas of fabrics in apparels generally include inner wear, fashion wear, functional wear, men's wear, ladies wear, and kids wear. These application areas can be further divided into many subcategories including but not limited to belts, coats, dresses, footwear, leggings, tights, gowns, headgear, jackets, neckwear, robes, cloaks, shawls, skirts, trousers, and shorts. These subcategories may be created on cultural, seasonal, and regional bases. The field of apparel fabrics is usually led by the fashion industry, and the most valued parameters include uniqueness and occasion-based clothing.

Figure 1.6 gives a comparison between the textile and apparel trades on global scale (fibre2fashion.com 2016). The term textile includes all textile auxiliaries from yarn to fabrics. It shows a clear inclination in the trade of apparels, worldwide. It is expected that the trade of apparels will be twice the trade value of textile items by the end of 2020.

Nonwoven fabrics are one of the oldest and simplest textiles. A classic example is felt. Earlier nonwoven fabrics were used in protective clothing, and shelters. In recent years, there is increase in nonwoven industry, and unconventional fabric trends are emerging, including geotextiles, diapers, bags, filters, etc. Now, nonwovens have begun to find applications in fashion in the clothing industry also. Research and development in the properties of nonwoven fabrics lead to this outcome of its use in apparels with improved properties. Unlike traditional fabric manufacturing process, where the fibers are converted to yarn and then woven, nonwoven fabrics are obtained directly from the fibers.

Recent research has led to production of fabrics with better draping, hand, durability, resilience, and recovery. Therefore, now these fabrics widely find their application in the insulation components of the garment interlining industry, clothing and gloves.

In woven fabric industry, the most used woven apparels are shirts, pants, coats, dresses, jackets, jeans, and blouses. In contrast, knitted apparels are mostly used for T-shirts, polo shirts, trousers, sweaters, undergarments, skirts, and children wear due to their inherent characteristics of stretch and comfort properties.

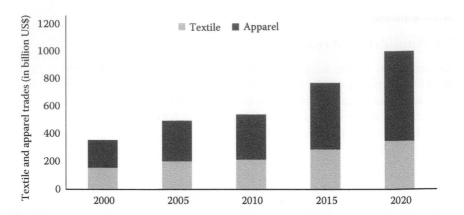

FIGURE 1.6
Comparison of textile and apparel trades. (Adapted from fibre2fashion.com. 2016. Indian Textile Industry—Prospects for the Next Decade. Accessed June 2. http://www.fibre2fashion. com/industry-article/6524/indiantextile-industry-prospects.)

1.4.2 Home Textiles

Home textile deals with the applications of textiles for household purposes. Home textiles are used mainly for their functional and esthetic properties aiming to provide comfort and mental relaxation to the people. The basic articles of home textile are grouped as sheets and pillow cases, blankets, towels, table cloths, and rugs. Both natural and man-made fibers are preferred in fabrics used for home textiles (Das 2010). This is an underexplored area, and there is a lot of scope for new developments and innovations.

These include a vast range of functional as well as decorative products used mainly for decorating our houses. Some of the most common home textile products are (Rowe 2009) home furnishing fabrics, bed spreads, blankets, pillows and pillow covers, cushion and cushion covers, carpets and rugs, wall hangings, different types of towels, table cloth and mats, kitchen linen and other kitchen accessories, bathroom accessories, etc.

The bed sheets and pillowcases are usually produced from plain woven fabrics either from cotton or from cotton/polyester-blended yarns. A small proportion of these are also produced from linen, silk, acetate, and nylon, with weave constructions varying from plain to satin, rarely knitted.

Blankets are usually woven using high-twist yarn in warp and soft (low-twist) yarns in weft. Other ways to produce are either by knitting or by flocking fibers. The composition and construction of yarn decide the degrees of warmth, softness, and durability of blanket. Thermal blankets are produced using a honeycomb weave or knitted to produce a lightweight open construction. The flocked polyurethane blankets have a polyurethane foam

base, which is covered with fiber flocking using an adhesive. Such blankets are very soft and spongy, and relatively light weight.

Terry towels are used to absorb moisture from wet skin, and therefore, they must be strong enough to withstand the stresses produced during rubbing, tugging, and laundering. These are woven piled structures produced either from cotton or from a blend of cotton and polyester yarn. Polyester provides increased strength and faster drying, while cotton provides greater absorbency.

Table cloths are produced in a number of designs and ways from cotton, linen, rayon, polyester, or their blends. The most popular constructions are damask and lace. The linen damask is the most expensive, and it requires greater care of laundering and ironing but offers advantages of luxuriousness and durability.

Floor coverings including rugs and carpets serve as a colorful foundation for the rooms, enhancing its beauty. In addition, they also act as the heat and sound insulators. The carpets may be produced by hand (woven) or by machine (tufted, woven, nonwoven, or knitted). The hand-woven carpets are very expensive; though their production rate is low, highly skilled labor is required for its production. Nonwoven carpets are produced from polypropylene (PP). The PP fibers are converted into web by needle punching, and thermal bonding technique is applied to produce carpet. They are the cheapest form of carpet and are not very long lasting.

1.4.3 Industrial/Technical Textiles

These are the textiles that are produced for certain functional properties rather than decorative purposes. The technical textiles are classified into 12 different segments as follows (Horrocks and Anand 2000): agrotech (for agriculture and aquaculture), buildtech (for building and construction), clothtech (for clothing and footwear), geotech (for civil engineering and geotextiles), hometech (for household textiles), indutech (for industrial applications), meditech (for medical and hygiene), oekotech (environment-friendly products), packtech (for packaging), protech (for protection), sporttech (for sports and leisure), and mobiltech (for automobiles).

The fabrics used for the technical purposes are mostly woven or nonwoven (Iqbal 2009). The knitted fabrics have a very low share in the industrial applications due to less stable structure as shown in Figure 1.7 (Lombaert 2014). The warp-knitted fabric structures have found some technical/industrial applications. The choice of fibrous material is largely dependent on the area of application, and it may include natural, man-made, and high-performance fibers.

Some of the examples for the technical use of fabrics are nets, ropes, jute bags, reinforcement for composites, tents, sewing threads, interlinings, waddings, geomembranes, shades, conveyor belts, hoses, filters, carpet backing, printed circuit boards, seals, gaskets, bandages, sutures, diapers, tea

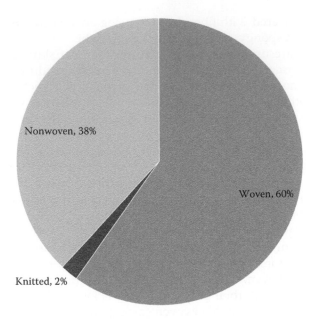

FIGURE 1.7
Woven, nonwoven, and knitted fabrics shares in technical textiles. (Adapted from Lombaert, F. 2014. In *Dubai Techtextil Symposium*.)

bags, electromagnetic shielding, ballistic protection, noise absorption, seat belts, flags, coated-inflatable life rafts, sleeping bag liner, poncho, surgical gowns, etc.

1.5 Global Textile Trade and Trends

The global textile and clothing trade, comprising all kinds of textile fibers, was around US$ 733.3 billion in 2012 (Hussain 2013). The different categories and their share are given in Table 1.1. It can be observed that the apparels and accessories as a whole (knitted and others) comprise about 55% of the global textile trade. This large share may be attributed to the value addition of the product. If we consider fabrics, the nonwovens, knitted, and woven fabrics (including yarn and fiber) account for 3%, 4%, and 9.5% shares of global textile trade, respectively.

Cotton, being the major natural fiber, has the highest production and consumption. Cotton fabrics have a number of applications ranging from apparel, home textiles to technical textiles. The cotton woven fabrics are exported in different forms (unbleached, bleached, dyed, yarn dyed, printed). The segregation of these categories is given in Table 1.2.

TABLE 1.1

Global Textile Trade

Category	Global Trade (US$, in thousands)	Share (%)
Silk fiber products	3,143,424	0.43
Wool fiber products	14,311,306	1.95
Cotton fiber, yarn, and woven fabrics	66,279,625	9.04
Plant fiber (excluding cotton) yarn, fabrics	3,760,447	0.51
Man-made filament products	45,581,281	6.22
Man-made staple fiber products	36,947,101	5.04
Nonwovens, waddings, felts, and chords	21,994,542	3.00
Carpets and floorcoverings	14,369,555	1.96
Special woven and tufted	12,432,921	1.70
Coated and laminated textiles	24,219,223	3.30
Knitted fabrics	29,999,926	4.09
Apparel and accessories (knits)	211,269,304	28.81
Apparel and accessories (nonknits)	193,872,398	26.44
Made-ups including home textiles	55,129,619	7.52
Total	733,310,672	

TABLE 1.2

Cotton Woven Fabrics Global Trade

Product Category	Global Trade (US$, in thousands)
Unbleached cotton woven fabric	5,816,747
Bleached cotton woven fabric	2,160,778
Solid dyed cotton woven fabric	7,813,446
Yarn dyed cotton woven fabric	4,281,063
Printed cotton woven fabric	5,124,538
Denim fabric	4,681,829
Total	29,878,401

References

Adanur, S. 2002. *Handbook of Weaving*. 1st edn. New York: CRC Press.

Balter, M. 2009. Clothes make the (hu)man. *Science* 325:1329, doi: 10.1300/J155v06n02_19.

Das, S. 2010. *Performance of Home Textiles*. New Delhi: Woodhead Publishing India.

fibre2fashion.com. 2016. Indian Textile Industry—Prospects for the Next Decade. Accessed June 2. http://www.fibre2fashion.com/industry-article/6524/indian-textile-industry-prospects.

Horrocks, A. R. and Anand, S. C. eds. 2000. *Handbook of Technical Textiles*. New York: Woodhead Publishing Ltd.

Hussain, T. 2013. State of textile and clothing exports from Pakistan. *TEXtalks* July/
    August: 62–65.
Iqbal, M. K. 2009. The applications of nonwovens in technical textiles. *Pakistan Textile
    Journal* 58(12): 35–39, doi: 10.1533/9781845699741.2.136.
Kvavadze, E., Ofer, B.-Y., Anna, B.-C., Elisabetta, B., Nino, J., Zinovi, M., and Tengiz,
    M. 2009. 30,000 Year old wild flax fibers. *Science* 325: 1359, doi: 10.1126/
    science.1175404.
Lombaert, F. 2014. Woven fabrics and their applications in technical segments pica-
    nol today: Key figures. In *Dubai Techtextil Symposium*, January 19–21, Dubai.
Nawab, Y. ed. 2016. Textile raw materials. In *Textile Engineering*. Berlin, Germany:
    De Gruyter.
Rowe, T. ed. 2009. *Interior Textiles*. Cambridge: Woodhead Publishing.
Shaker, K., Munir, A., Madeha, J., Salma, S., Yasir, N., Jasim, Z., and Abdur, R. 2016.
    Bioactive woven flax-based composites: Development and characterisation.
    Journal of Industrial Textiles 46(2): 549–561, doi: 10.1177/1528083715591579.
Wulfhorst, B., Oliver, M., Markus, O., Alexander, B., and Klaus-Peter, W. 2006. *Textile
    Technology*. Munich, Germany: Hanser Publishers.

2

CAD for Textile Fabrics

Khubab Shaker and Haritham Khan

CONTENTS

The introduction of computers to every walk of life has widely influenced the human life, making it easier and comfortable. The computers are serving to design, analyze, and manufacture the products without any wastage of materials, in a short span of time, thus increasing the productivity [1,2]. Computer is an electronic machine capable of performing logical and mathematical calculations and processing tasks in accordance with a prewritten set of instructions. This detailed set of instructions is termed as a computer program. It is a fast and accurate data-manipulating system that is designed to automatically accept and store the input data.

The application areas of the computer may be classified broadly into engineering, business, scientific, and recreational applications. The computers are widely used in the engineering applications to help both designing and

manufacturing industries manage their processes and information system. The role of computer in manufacturing is broadly classified into two groups, namely design and process control. The focus of this chapter is on the engineering application areas of computers, especially in design.

2.1 What Is CAD?

The term CAD stands for computer-aided design and is defined as the use of information technology in the design process. The CAD system is usually a combination of hardware and specifically designed software. The hardware includes graphic devices and their peripherals for input and output operations, while the software package manipulates or analyzes shapes according to user interaction. The core of a CAD system is its software, which takes input from product designer, makes use of graphics for storing the product model, and drives the peripherals for product presentation. Use of CAD does not change the nature of design process but aids the product designer. Most of the CAD software packages have the provision to create three-dimensional (3D) models that can be rotated and viewed from any angle. These state-of-the-art modeling CAD packages help architects, engineers, and designers in their design activities.

2.2 Why CAD?

The CAD facilitates the product designer by providing the following [3]:

- Graphical representation of product
- Easy modification of graphical representation
- Record the whole design and processing history
- Complex design analysis using finite element method

The graphical representation is so accurately generated that the user can nearly view the actual product on the screen. The design is easily modifiable, and it helps the designer to present his ideas on screen without any prototype. In addition, using certain analysis tools (e.g., finite element methods), the user can perform mechanical, heat transfer, fluid flow or motion analysis, etc. for design optimization.

The CAD systems help to reduce the design and development time of a particular product. Figure 2.1 provides a representation of the generalized

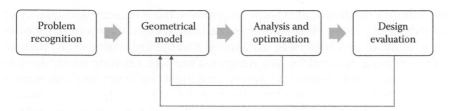

FIGURE 2.1
A generalized traditional design process.

design process. Shorter the development time of a product, earlier it can be introduced in the market, and the product may have an extended life span if the built-in quality of it is up to the mark. In addition, the CAD systems also enable the application of concurrent engineering and have substantial impact on the cost, functionality, and quality of final product [4].

One of the initial applications of CAD was for 2D drafting, but the current focus is on the 3D solid and illustration of the real part. The CAD packages have the ability to model the complete assemblies and analyze a virtual prototype for its performance. The 3D illustration can also be exported to other analysis/representation programs, and it can serve as a communication medium between people from different departments of an organization.

2.3 CAD for Weaving

Textiles are structurally complex materials and therefore represent ongoing challenges in the product design and component selection. The component selection involves the selection of material (fibers), yarn type, and fabric geometry and construction. A lot of research is ongoing in this area, and now it is possible to analyze the complex textile structures and optimize their desired properties. The engineering analysis tools such as finite element modeling (FEM) allow the analysis of material behavior without constructing actual models of the part being designed. FEM involves the graphical division of structure (domain) being analyzed into small portions called elements. The interrelated response of these elements to force is then determined by computer program using a predefined set of equations. The graphical representation of results helps the designer to observe the areas where deformations and strains occur. It allows the designer to modify the design at those particular points and analyze the modified design [5].

The CAD systems dedicated to textiles industry offer solutions that not only help in the design stage but also in the manufacturing process, bridging the gap between design and the production stages. This electronic communication between the design package and the production tool (weaving

and knitting machines, etc.) helps to improve the efficiency and quality of the industrial process. The vibrant textile industry has always been innovating in terms of new materials, machinery, and the processing cycle, with an ever-increasing demand for new designs. Therefore, creating newer designs for textile and fashion articles has become a great challenge for the designers than ever before and necessitated the use of CAD packages. The manufacturer can have regular interactions with the customer and make frequent and quick changes to the design, meeting the latest trends in textile and fashion industry. It not only helps to reduce the pressure on manufacturer but also provides the customer satisfaction, with improved designs.

The textile designers were more confined to the fabric design and did not correlated the yarn and fabric characteristics. Now, the entire fabric design process has been revolutionized incorporating yarn characteristics into the design process. Previously, the designers had laborious work on the graph paper and stencil, which is now simplified using a mouse or stylus pen and computer to produce innovative designs. Introduction of the CAD technology through implementation of designs and color combinations is making the textile fabrics more striking and viable. It is also helping to meet the quick-changing expectations of the consumers for fashion designs.

Woven fabrics are the most thoroughly investigated textile structures. A lot of research is in progress in this domain. The use of CAD for woven fabrics is highly dependent on the end use of fabric. The two most important groups of woven fabric simulation are as follows [6]:

- CAD for esthetic or artistic design
- CAD for the woven fabric structure/geometry

Both of these categories have their own importance in different fields. The aesthetic or artistic design perspective is of utmost importance for the fabric designers. They use different color combinations of yarn along with linear density, thread density, weave, etc. to get a particular appearance. In contrast, the 3D simulation of woven fabric structure/geometry is further used by mechanician in engineering calculations (mechanical, thermal, etc.) as well as for the artistic design.

The different CAD packages used for weaving are listed in Table 2.1. These CAD packages make use of different modules for the graphical representation of the fabric structure and appearance.

2.4 CAD for Esthetic or Artistic Design

It is the computer-aided designing from a designer's perspective, focusing on the esthetic/artistic appearance of fabric. These simulations are probably

TABLE 2.1

Common CAD Software Packages for Woven Fabrics

Software	Modules
Nedgraphics	Dobby/Jacquard/Textile Designer/2D or 3D geometry
Point Carre	Dobby/Jacquard/Textile Designer
Scotweave	Dobby/Yarn/Jacquard/2D or 3D geometry
WeaveMaker	Dobby
Weave Point	Dobby
YXENDIS	Dobby
Arhane	Dobby/Jacquard
PENELOPE	Dobby/Jacquard/Terry
Textronics	Dobby/Jacquard/Carpet
Texgen	2D/Multilayer/3D geometry
DB weave	Dobby
EAT (Design Scope Company)	Dobby/Jacquard/Multilayer/3D/Knitting
WiseTex	Geometrical model of fabrics (woven, knitted, laminates, braided)

the earliest commercially available simulations of woven structures, developed to replace the sophisticated work of designers, preparing punch cards for jacquard machines. These CAD systems permit the user to visualize the pattern in various fabrics and alter its appearance in a user-friendly (i.e., usable without extensive knowledge) environment. All of these simulations start from the weave pattern definition since it contains the information about the appearance of the yarns from both sides.

This principle is being used by most computer programs to simulate the appearance of the woven structure, without re-constructing the real appearance of the fabric. Using pictures of real or created yarns, appropriately placed and rotated, it is possible to present a realistic view of the structure. Nowadays, most programs provide a visual representation of the fabric.

Within a selected range, various fiber and yarn attributes and weave and knit characteristics can be tested in computer visualization without the expense of producing trial fabrics for all of the considered combinations. The design can be viewed on high-resolution monitors, and using separate flat screen displays, high-quality photographs can be produced. Copies of the displays can also be produced on plotters, with the resolution and appearance dictated by the plotter quality. Many of the available programs can be run on microcomputers, reducing the cost considerably.

The major modules involved for the fabric appearance include Dobby, Jacquard, and Terry. In addition, there are some ancillary modules helping to simulate the fabric, comprising of the yarn, color, and weave library [7]. Information about these modules and libraries has been provided in brief below.

2.4.1 Color Library

The aim of this module in the CAD software is to enable the user create a library of colors according to his own choice. This module also helps to create, save, and load the color pallets when required. In addition, the color specifications are also available, which allow the user to edit a certain color using its specifications. More the number of colors in the library, easy it will be for the user to view the fabric in different color combinations. The different CAD packages have a varying number of colors in its library, depending on the specificity of its end use and area of application. Using the color library helps to view the simulation of fabric in 16.7 million color combinations.

2.4.2 Yarn Library

Various kinds of yarns (either regular or fancy) are used in the textile industry, and the yarn library stores these yarns for fabric simulation. It also offers the facility of yarn creation, which retains all of the technical details of a particular yarn. The user can specify the yarn fineness/linear density (in terms of thickness or tex/denier system), twist direction (S or Z), yarn color (from color library), etc. in the yarn information dialog box. This particular yarn can also be stored for fabric simulation at a later stage.

Some basic CAD systems make use of colored rectangles as yarn to give the fabric simulation. The width and height of these rectangles help to specify the yarn thickness and yarn spacing. Such systems give only a rough appearance of the fabric, not taking into account the yarn imperfections. Other CAD systems have a library of scanned yarns (taking into account the imperfections) or have the capability to scan a particular yarn and use it for the simulation.

2.4.3 Weave Library

The woven fabrics are constructed by interlacing two sets of yarns (warp and weft) perpendicular to each other. The interlacement pattern known as weave can vary greatly, affecting the fabric geometry and appearance [8]. The weaves are usually classified into basic weaves, derived weaves, combined weaves, and complex weaves. In addition to these, woven fabrics having considerable thickness are also produced from multiple sets of yarns in each direction (called multilayer woven fabrics). Weave is the starting point for CAD of a particular fabric as it contains information about the appearance of yarns from both sides.

The usual weave patterns for the single-layered structures are represented by a rectangular array, having cross marks at the interlacement point when the warp yarn lies above the weft yarn. While representing the weave in matrix form (2D binary weave matrix), this mark is replaced by the digit "1" as shown in Figure 2.2. Here, "1" shows that the warp yarn lies above the weft yarn, while "0" shows the weft yarn lies above the warp yarn. This coding is

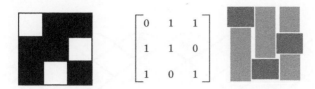

FIGURE 2.2
Pointed paper, matrix, and fabric forms of weave.

used to the control heald frames or the Jacquard unit of the weaving machine, as well as for the reconstruction of the geometry of the woven fabric.

This principle was and is still used by most computer programs to simulate the appearance of woven structure, without having to reconstruct the real appearance of the fabric. Using pictures of real or created yarns, appropriately placed and rotated, it is possible to present a realistic view of the structure. This method of representation is suitable for the basic as well as derived and combined weaves. Complex weaves are mostly used for the 3D multilayer woven fabrics. These fabrics are characterized by their structural integrity and notable dimension in the thickness direction [9] and are classified into multilayer, orthogonal, and angle interlock structures. The 3D multilayer fabrics have multiple layers of warp and weft, and the interlacement pattern is not as simple as it appears to be in the case of basic weaves. The matrix representation of such a fabric structure is based on the levels as shown in Figure 2.3. More details of complex weave structures are provided in the later sections.

The fabric simulation is the virtual fabric produced of a certain design before the production of actual fabric. Simulation helps the user to make any amendments to the weave design, yarn specifications (thickness, type, density), or color combination to get the desired appearance. The fabric simulation and actual fabric are almost alike, giving authenticity and reliability to the CAD process. The great quality of the simulations reduces the costs of samples and economizes the time of response to the clients. The CAD packages use different modules of fabric simulation based on the fabric appearance and properties.

2.4.4 Dobby Module

This module is dedicated to the dobby industry. The dobby is a shedding system, and looms with this system are designated as dobby looms, irrespective of the picking media. Dobby is a relatively complex shedding system, and it controls up to 32 heald shafts. The old systems of providing lifting plan/peg plan included peg chain, peg cylinder, punched papers, and plastic pattern cards. All of these techniques had a limited pick repeat, and dobby fabrics having a small repeat could be produced. Currently, the pick repeat is provided by the computer programs (dobby module) and is virtually unlimited.

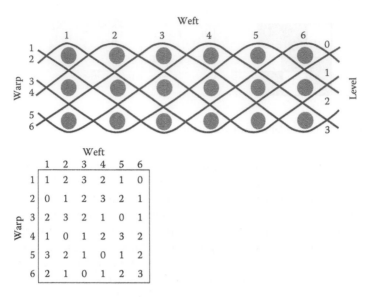

FIGURE 2.3
Structure of a three-layered angle interlock structure and code for WiseTex.

This system offers more design possibilities as compared to the old dobby-shedding systems [10].

Dobby has a collection of useful tools for easy creation of any kind of weave. It helps in the automatic generation of draft and peg plan from a given weave or vice versa. The graphical user interface (GUI) allows the user to work on all of the three views (weave design, draft, and peg plan) simultaneously. The module also helps the user to either create a weave or generate a peg plan, keeping the drawing in draft (DID) constant. Changing peg plan only, a wide range of different weave designs can be created effortlessly, without changing the drafting order repeatedly. It is highly advantageous for the sampling purposes, where a number of fabrics can be produced by just changing the lifting plan.

Some of the dobby modules are capable of creating specialty fabric simulations, for example, seersucker, double fabrics, and double-face fabrics. Seersucker fabric has a wavy appearance due to differential warp/weft tension. The system offers the possibility of modifying different fabric parameters of simulation. On the other hand, it allows to apply dyeing effects on fabrics, with the requested colors. The simulations are of great quality and realism, due to the facility of fancy and regular yarn creation. Combined with fantastic finishing effects that can be applied to the fabric, spectacular results are achieved. Other specifications of this module are as follows:

- Color management (applying warp color pattern to the weft automatically, interchanging of warp and weft color patterns)

- Fabric simulation with different denting patterns
- Reverse, mirror, or invert the fabric appearance
- Assignment of different weaves in the same fabric (merging two designs)
- Export the fabric in "jpeg," "bmp," or "tiff" format

2.4.5 Jacquard Module

The jacquard design is a unique combination of artwork, weave, and yarn specifications. Most of the jacquard designs start with the artwork designer, which is an image creation tool for the preparation of artwork for use in the jacquard design. The basic artwork can be created freehand, scanned into the artwork, or loaded from an existing file. Freehand artwork creation involves the freehand painting directly onto the computer screen using a pencil or any drawing tools. An existing paper artwork can be scanned onto the computer screen using a color scanner. Once on screen, user can use the freehand artwork tools to manipulate and edit this image [11]. Artwork designer can also import and export most of the graphics file formats; therefore, image may be created and exported using other graphics software (Photoshop, Corel Draw, etc.).

Weave structures are created for each unique color in the graphics image, generally different from those around it [12]. The weave structures are applied to the graphics image to create a single total weave structure. Warp and weft patterns are created and applied to the total weave structure to generate the woven jacquard design. In order to keep the number of weaves to a workable minimum number, the number of unique colors must be controlled in the artwork. Single pixels and single pixel width lines are avoided in the jacquard artwork for the clarity of design.

The color reduction/compression is necessary when there are too many colors in artwork image and user wishes to reduce the number of colors (and hence the number of weaves). Physical appearance of colors in artwork helps to distinguish between different color areas. The colors in finished design come from warp and weft patterns and not from the artwork. Thus, in addition to the image, the jacquard designer program uses extra information stored alongside the image file, which standard graphics file formats do not include. For this reason, the user must go through the Artwork Designer program as part of the jacquard creation process.

The jacquard design module provides a wide range of possibilities such as designing tools, simplifying the process, and obtaining real simulations of the fabrics that will be woven later. The software helps to generate the loom file that contains loom layout, according to the machinery with which the company works. Therefore, it is possible to save different loom layouts and reuse them for later works, which saves time in future designs. Once the loom layout has been realized, the looms can be connected directly to

the management system of the company. The user can then export the information for direct connection to the looms or management systems. In the same way, the CAD system can create the manufacturer order to control the production. The major realizable functions of jacquard design are as follows:

- Color reduction in artwork
- Elimination of spots
- Wide range of drawing tools
- Elimination of floats
- Visualizing different repeat compositions in the same image
- Weft control according to design

2.4.6 Terry Module

The piled fabrics are produced by loops or strands of yarn (termed as pile) raised over the surface of a fabric. Some common examples of piled fabrics are carpets, velvet, corduroy, and terry towels. The textile and fashion companies are developing new tools that allow the designers to work easier and faster. With the incorporation of the software into the current process of loop fabric design, the following advantages are obtained:

- Capture images with the scanner or a digital camera
- Filter the colors in image and improve the image quality
- Create multidensity graphs and define drawing with a different weave for every color
- Possibility to work with layers
- Facility at conversion of images to a graph
- Assignment of elementary weaves in a multidensity graph
- Simulation of a terry fabric in high quality
- Define the length and thickness of the loop
- If looms connected to a network, send design directly to loom or management system

2.5 CAD for the Woven Fabric Structure/Geometry

Nowadays, textile fabrics are used extensively in technical applications. In such application areas, precision in structural design and its properties (behavior of the fabric) are of primary importance rather than only esthetics.

The textile-based composite materials are widely used in advanced applications such as aircraft. Thus, the behavior of fabric structure is vital in achieving the overall aircraft efficiency and safety. The performance of fabric is basically a function of the geometrical construction of yarns in it. The behavior of textiles under different loading conditions, whether mechanical, thermal, or fluid, is of utmost importance for the designer.

Geometrical models of woven fabrics allow the estimation of some structural and physical properties of the fabrics, for example, areal mass and porosity. The geometrical models of textile assemblies are therefore more important, providing reliable geometrical information on textile assemblies for finite element analysis (FEA) for performance simulation.

The initial model of the fabric geometry is supposed to be presented by Peirce on plain fabric in 1937 [13]. The model was based on a number of assumptions like circular yarn cross section, complete flexibility of yarns, incompressible yarns, and arc–line–arc yarn path. A set of eight equations were established to describe the geometry of plain woven fabrics. These equations were based on 13 variables (modular heights, crimp, diameter, thread spacing, modular length, weaving angles of warp and weft yarns, and sum of the diameters of warp and weft yarns). Therefore, with five variables known, these simultaneous equations can be solved for definite fabric geometry.

Peirce's models were further extended by the researchers, addressing the assumptions of his model. Kemp's racetrack model (1958) and Shanahan and Hearle's lenticular model (1978) are extensions to Peirce's model, in that the circle (arc) and straight-line elements are adopted, although the yarn flattening is satisfactorily addressed. In each case, the models are used to determine the maximal warp and weft densities in the fabric using the jammed fabric theory. According to the Peirce model, the straight-line segment in the warp yarn path decreases when the weft density increases and disappears at the maximum weft density [14].

In order to create CAD for such fabrics, mathematical expressions of their weaves are also essential. It simulates the entire yarn geometry in 3D space. Such models are based on the description of yarn path on its plane. The cross section of weft yarns is shown in the plane of warp yarn path. It allows the definition of contact points and types of lines between them. The simplest model is based on circular cross section of yarn and modeling yarn path using circular arcs and lines (Peirce, 1937). The yarn axis is calculated using the set of coordinates obtained from the following equation:

$$S_j = \left[(x_1, y_1, z_1),\ (x_2, y_2, z_2), \ldots, (x_n, y_n, z_n) \right]$$

where the term (x_1, y_1, z_1) represents the positions along x, y, and z axes for the first yarn. Around yarn axis, volume is built and yarn cross section is calculated. The fabric geometry is used for the FEA to simulate different behaviors like mechanical, thermal, fluid dynamics, etc.

2.6 CAD for Knitting

There is a thousand years old history of knitwear. With the introduction of the computer, simulation of knitwear emerged as a promising field for the researchers. Simulating a knitwear is more efficient and more realistic approach for new developments. Thus, hit and trials for production of samples on the machine is eliminated, which helped to save time and money. It also makes possible to minimise errors, and the design can be transferred by a single click from one computer to another. Simulation of knitted fabrics therefore became a major research interest in recent years [15]. Applications of CAD in knitting can be broadly categorized as follows:

1. Knitting structures design
2. Design of knitting pattern shapes

In first technique, knitting structures design, some standard construction are used to make conventional designs like the model of jacket, sweater, etc. Here, construction means whole style of the desired end product, including shape and sizes. The CAD has facilitated the designers to make use of different knitting techniques including jacquard, intarsia, gusset and lace in an effective manner. While second technique, knitting pattern shapes, is capacity of the machine to make fully fashioned products. The machine knits and cuts the product. Knitting machine knits the whole products to avoid separate cutting and sewing. This will help reduce work in progress stations (WIPS) and number of operations resulting reduction in material wastes [16].

2.6.1 Knitting Structures Designs

A CAD software used to produce knitted structures is provided below.

1. Starfish Software

It is a simulating software that works on a principle that, in order to produce a knitted fabric with desired dimensions and performance, it must be first known that what type of finished product it would be.

After several laboratory- and industry-level preliminary testing, the STARFISH kit was commercialized for the first time in 1988. Collection of new data and development of analytical techniques in it are continued by Cotton Technology International (CTI).

For any finished product, the end consumer can/may be an individual or an organization for which the performance of the product matters a lot.

Customer presents his requirement in the form of a specs sheet in which all details of the product are mentioned, including weight, width, shrinkage,

pattern, etc. In some cases, especially when the product is new and novel, it undergoes many changes and keeps on changing because of rising demands and opportunities in market. To meet the diverse customer needs, the manufacturer has to make a number of trials, which may be costly and time consuming. The STARFISH provides the opportunity to check the feasibility of intended samples (production parameters). This not only helps to reduce sample cost but also to give quick and easy response to boost up customer satisfaction.

2.6.2 Design of Knitting Pattern Shapes

Some forms of CAD are also used in designing knitting pattern shapes. Some of the CAD forms/softwares related to knitting are given below.

1. ProCad

In multibar and Jacquard-Raschel, ProCad is used as a perfect tool to produce patterns. It is a user-friendly and very effective interface used in today's professional ateliers. Karl Mayer GMBH and TEXION are supplying this software, which is compatible with almost all machines supplied by Karl Mayer and LIBA. Since its introduction, it is regularly upgraded to a new and challenging form to meet rising customer needs. There are different modules of ProCad with different functions. Some of the modules are ProCad developer, ProCad simulace, ProCad Simujac, ProCad velours, ProCad warp knit 3D, and ProCad knit assistant. These modules help the user to operate on almost all knitting machines including Multibar, Jacquard electronics, Tricot and velour designs on Karl Mayer and LIBA knitting machines. Scanned design drafts or CAD data that can be converted into designs are also imported through the ProCad program.

Karl Mayer has recently developed a software known as Texion's Procad warpknit 3D, which is unique of its type, and it helps warp-knitting machine manufacturers in many ways. Detail of the module is not in the scope of this book.

2. ProFab

This is considered as the intelligent software solution for the networking of textile knitting machines. ProFab, introduced by TEXON, is a new milestone in the history of automation field. In common practice, involvement of human being and flow of information from one to another and their engagement in more than one activity at a time have resulted in machine manufacturers to integrate software within the machines to reduce human involvement to a very limited extend. ProFab provides TEXION manufactures with high degree of automation.

ProFab is characterized by a modular structure consisting of four elements:

- Design manager, which is a software with a solution for secure design data transfer
- Job manager, which is a software that gives optimal results in advanced planning and scheduling
- Machine manager, provides solution for monitoring production processes
- Beam manager, is related to management and it gives solutions related to warp beam management

Thanks to its open structure, ProFab Network can be enhanced with the addition of extra system functions, but without this influencing its other elements. This makes ProFab Network a safe, long-term bet.

3. Shima One

Sheima Sheiki has introduced Shima One to design flat-knitting patterns, intarsia patterns, and jacquard patterns. It can be used in simulation of mohair, slub, and shiny yarns taking into consideration the plating number of thread take-ups and counts. Sheima One also improves production efficiency by using database of more than 1000 patterns, and these patterns are connected with Knit CAD software for knitting machines (Shima Seiki). SDS-ONE APEX3 not only helps designing and linking these patterns with knitting machine, but it also helps each stage in flat-knitting machine starting from knitting pattern, pattern design, and colorway evaluation to production and sales promotion. It is a feasible and more friendly system of knitting production not only for manufacturers but also for planning companies, OEM/ODS manufacturers, trade firms, yarn traders, interior designers, and sundry manufacturers. This CAD system is also used in circular knitting machines for stripes jacquard patterns, single-knit fabrics including jersey and tuck patterns, and double-knit fabrics such as rib fabrics.

2.6.3 Limitation of CAD

Although CAD systems are used to make machines or products user friendly, increase production, and get right thing at right time, it has some disadvantages as well.

Its major disadvantages are high capital cost and highly skilled and trained operators or designers.

2.6.4 Advancements in Knitting

CAD has made the life of textile engineers easy, and it has added a new taste to the technology of knitting. Competitive edges given to knitting because of the addition of computer are given below.

2.6.4.1 Computerized Production

Computer Aided Designing (CAD) and Computer Aided Manufacturing (CAM) were first introduced in 1970, and since then, CAD is used by designers to create product design, and the designs are transferred to CAM machines to manufacture the final product. These computer-aided tools or systems have replaced old mechanical shaping and patterning devices on machines.

This replacement has made possible quick response to changing demands, and this has made possible developing new designs through CAD and converting them on machine through CAM. In early days, these systems were expensive, where only the major companies could afford them, but later their prices fell, which made even small- and medium-sized companies to invest in this new technology. These new computerized technologies have enabled companies to work globally and made everything online, where a person sitting in his office in one corner of the world can view production in another corner. In this way, designs produced in one part of the world can be transferred to another part electronically to low-cost producers [17].

2.6.4.2 Whole Garment Knitting (Seamless Garment)

Whole garment knitting or seamless knitting is a new and novel concept in knitting, which has distinguished features. Cutting knitted fabric into pieces of different shapes and sewing them to create garments is an old and time-consuming technique. Since the 1970s, companies are working to introduce new technologies that could produce a complete garment in one step. In the 1970s, it was made possible to manufacture garments in one process without the loss of fabric while cutting and sewing. This new concept was modernized and upgraded to whole garment knitting in 1980 and Sheima Seiki made first whole garment knitting machine in 1990. This advancement is due to the introduction of new needles called slide needles by Sheima Seiki. These needles consist of hooks located centrally between a flexible two piece slider mechanisms. The sliders help transfer stitches during the knitting process. Whole garments having no or low seams are more comfortable than sewn clothes. No or low seams means that the fabric can be fit to the body and stretch more easily. Whole garment machines are also capable of producing complex designs, and 2D fabrics are replaced by 3D knitwear, shaped and pleated fabrics [18].

References

1. A. Dwivedi and A. Dwivedi, Role of computer and automation in design and manufacturing for mechanical and textile industries: CAD/CAM, *Int. J. Innov. Technol. Explor. Eng.*, 3(3), 174–181, 2013.

2. B. J. Collier and J. R. Collier, CAD/CAM in the textile and apparel industry, *Cloth. Text. Res. J.*, 8(3), 7–13, 1990.
3. N. Bilalis, *Computer Aided Design—CAD*, 2000.
4. Z. Stjepanovic, Computer aided processes in garment production: Features of CAD/CAM hardware, *Int. J. Cloth. Sci. Technol.*, 7(2–3), 81–88, 1995.
5. V. Vestby, CAD-CAM development in art and craft weaving: Experiences from integration in teaching and training in vocational and university level schools in Norway, *Educ. Comput.*, 6(1–2), 147–152, 1990.
6. D. Veit, ed., *Simulation in Textile Technology*. Oxford: Woodhead Publishing, 2012.
7. G. Goel, Implementation of CAD/CAM in weaving system, Fibre2Fashion, 2009. http://www.fibre2fashion.com/industry-article/4159/implementation-of-cad-cam-in-weaving-system?page=1
8. S. Kim, Development of a parametric design method for various woven fabric structures, *J. Eng. Fiber. Fabr.*, 6(4), 34–38, 2011.
9. K. Shaker, Y. Nawab, M. U. Javaid, M. Umair, and M. Maqsood, Development of 3D woven fabric based pressure switch, *Autex Res. J.*, 15(2), 148–152, 2015.
10. K. Mathur and A. M. Seyam, Color and weave relationship in woven fabrics, In: S. Vassiliadis, ed. *Advances in Modern Woven Fabrics Technology*. Rijeka, Croatia: InTech, pp. 129–151, 2011.
11. A. Mitra, CAD/CAM solution for textile industry an overview, *Int. J. Curr. Res. Acad. Rev.*, 2(6), 41–50, 2014.
12. S. Kovačević and I. Schwarz, Weaving complex patterns — from weaving looms to weaving machines, In: C. Volosencu, ed. *Cutting Edge Research in Technologies*. Rijeka, Croatia: InTech, pp. 93–111, 2015.
13. S. J. Kim and H. A. Kim, Data base system on the fabric structural design and mechanical property of woven fabric, In: P. D. Dubrovski, ed. *Woven Fabric Engineering*. Rijeka, Croatia: InTech, pp. 169–194, 2010.
14. J. Hu, ed., *Computer Technology for Textiles and Apparel*. Oxford: Woodhead Publishing, 2011.
15. X. Li, G. Jiang, and P. Ma, Computer-aided design method of warp-knitted jacquard spacer fabrics, *Autex Res. J.*, 16(2), 51–56, 2016.
16. E. I. Zaharieva-stoyanova, Methods of graphic representation of curves in CAD systems in knitting industry, In *XLII International Scientific Conference on Information, Comunications and Energy Systems and Technologies*, Ohrid, Macedonia, pp. 24–27, June 2007.
17. M. Neves and M. Assis, Geometrical and mechanical properties of jacquard and intarsia knitted fabrics, *Ind. J. Fibre Textile Res.* 19(September), 177–184, 1994.
18. M. Blaga, D. Dan, R. Ciobanu, and D. Ionesi, Interactive application for computer aided design of 3D knitted fabrics, *The 7th International Scientific Conference, eLSE—eLearning and Software for Education*, Bucharest, April 28–29, 2011.

Section I

Woven Fabric Structures

3

Introduction to Weaving

Syed Talha Ali Hamdani

CONTENTS

3.1 Introduction

Weaving is defined as a process of interlacing of warp and weft yarns at right angle to each other. There are practically an endless number of ways of interlacing warp and weft yarns. The warp yarn are sometimes termed as ends, and weft yarn are termed as pick or filling. In order to be the fabric as a single entity,

the interlaced yarns should have enough cohesive forces between each other, and woven fabrics must possess extra high length or width ratio to thickness. The process of weaving yarns into fabric, as shown in Figure 3.1, is performed on a "weaving machine," which is also called a "loom." The weaving process involves preparation of warp and weft yarns. The weft yarns that comes directly from spinning is used as it is, but sometimes they are rewound in small bobbins on winding machine to produce a larger weft yarn package. The warp yarn goes thorough warping and sizing before it is woven. After weaving, the woven fabrics are sent to folding department for inspection and grading.

In practice, the weaving machines are named after their weft insertion systems. Mainly, there are two main weft insertion systems, namely shuttle weft insertion and shuttleless weft insertion systems. Shuttle weft insertion is performed on shuttle looms, which have been used for centuries to make woven fabrics. In this type of loom, a shuttle, which carries the weft yarn wound on a quill, is transported from one side to the other and back. In the mid-twentieth century, shuttleless weft insertion started to emerge that used other forms of weft insertion mechanisms such as air, projectile, rapier, and water. Today, the shuttle looms have become obsolete and are not manufactured anymore except for some very special niche markets. The types of weaving machines are discussed in detail in Section 3.3.

The existing shuttle looms have been replaced by the shuttleless weaving machines in industrialized countries. Nevertheless, approximately

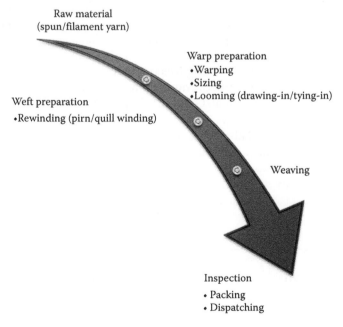

FIGURE 3.1
Process flow of weaving.

2.6 million of the 3.2 million looms existed throughout the world in 1998 were still shuttle looms. However, they are being replaced by the shuttle-less weaving machines at a fast rate. Today, the three most popular weaving machines are air-jet, rapier, and projectile machines [1].

3.2 Basic Weaving Mechanism

As defined earlier, the weaving process requires interlacing of warp and weft yarns at right angle to each other. In order to interlace these yarns, basic mechanism involves primary and secondary motions. The primary motion includes shedding, picking, and beat-up, whereas the secondary motions are warp let-off and cloth take-up.

3.2.1 Primary Motions

The motions that are compulsory for weaving process are called primary motions. Weaving will not happen if any of these motions are not completed. These motions include shedding, picking, and beat-up. The primary weaving motions are shown in Figure 3.2.

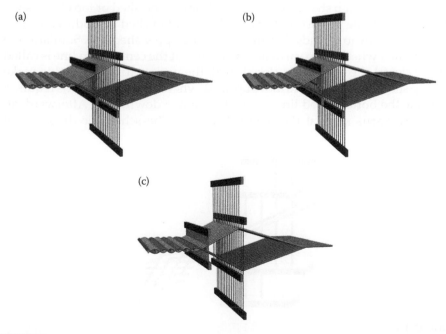

FIGURE 3.2
Primary weaving motions. (a) Shedding (b) Picking (c) Beat-up.

3.2.1.1 Shedding

This is a process of raising and lowering warp yarns by harnesses to make an opening for the filling (weft) yarn to pass through. In shedding motion, warp threads are divided into two layers. The top layer is called top shed line, and the bottom layer is called bottom shed line. The raised and lowered form of warp yarns is called shed, and there are three types of shedding motions available for different types of fabrics, namely tappet shedding [2], dobby shedding, and jacquard shedding. The shedding is achieved by means of treadles, dobby, or jacquard. The treadles are used in handlooms, operated by the weaver's feet, and in power looms, operated by shedding tappets. The dobby and jacquard are either mechanically controlled or electrically controlled shedding systems. Healds are used in tappet and dobby shedding systems, whereas jacquard controls the warp threads individually for producing sheds by means of hooks, needles, harness cord, and knives. A simple shedding motion controlled by harness is shown in Figure 3.3. On the basis of shed geometry, the shedding is broadly divided into two classes: closed shedding and open shedding.

3.2.1.2 Closed Shedding

The closed shedding system employs all of the warp yarn levels after the insertion of each pick. The level is made either at bottom/top or at the center of shed line. The type of closed shed where the level of warp yarns is made at bottom/top shed line is called bottom closed shed or top closed shed depending on the position of leveling. This kind of shed is produced by giving motion only to threads that are to form the upper shed line. Similarly, the type of shed where warp yarns are made level at the center shed line is called center closed shed. In center closed shed, the warp yarns required to make the top shed line are made to move upward, whereas the warp yarns required to make the bottom shed line are made to move downward. Afterward, all the warp yarns meet at the center shed line. The schematic diagrams of

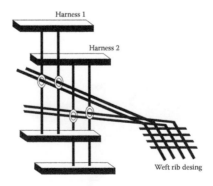

FIGURE 3.3
Shedding motion. (Reproduced from S. Adanur, *Handbook of Weaving*. Lancaster, PA: CRC Press, 2000. With permission.)

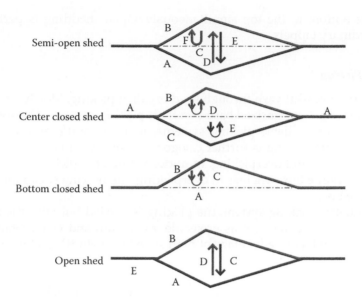

FIGURE 3.4
Semi-open, center closed, bottom closed, and open shed. (Textile Store. Reprinted with permission.)

bottom closed shed and center closed shed are shown in Figure 3.4. The advantage of bottom closed shed is to achieve high cover factor at the cost of high power consumption and wear and tear of weaving parts. The bottom/top closed shed is not suitable for high-speed weaving due to larger time required for changing the shed. The high-speed weaving can be achieved by center closed shed due to less strain in warp yarns as compared to bottom/top closed shed. The power consumption and wear and tear are also less in center closed shed as compared to bottom/top closed shed.

3.2.1.3 Open Shedding

In open shedding, the warp is only moved when a pattern requires a change of position. There are two methods of producing open shedding, that is, open shedding and semi-open shedding. In semi-open shedding, as shown in Figure 3.4, the stationary bottom line is retained, but warp yarns of the top shed line is either lowered to the bottom at one movement or raised to the top. The remaining warp yarns move down. This is formed under both open and closed principles and is being used by double-lift dobby and Jacquard shedding system.

In open type of shedding, as shown in Figure 3.4, the warp threads form two stationary lines, one at the top and the other at the bottom. After inserting a pick, threads are moved from one fixed line to the other. So, one line of thread is lowered from the top to the bottom, and the other line was raised

from the bottom to the top simultaneously. Open shedding is performed using ordinary tappets.

3.2.1.4 Picking

The insertion of weft yarn through shed is called picking. Mostly, the weaving machines are categorized based on their picking systems. There are two major types of available picking systems, namely shuttle and shuttleless picking. Shuttle picking is further categorized into two main systems, that is, underpicking and overpicking. In underpicking, the picking stick moves under the shuttle box, whereas in overpicking, the picking stick moves over the shuttle box.

In shuttleless picking system, the picking is carried out with the help of various picking media such as projectile, rapier, air, and water. Shuttleless picking system has an advantage of high speed over shuttle picking system. A number of weft (filling) selections are made available on weaving loom to select the desired weft depending on the count and color of weft yarn. A weft, being inserted through a shed, is shown in Figure 3.5.

3.2.1.5 Beat-Up

The filling insertion system cannot fit the weft at an acute angle of shed opening, which is done with the help of beat-up motion. The fitting of newly inserted pick to the fell of cloth is called beat-up. The fell of cloth is an imaginary line which shows the point of cloth woven. The beat-up is performed with the help of a device called reed. The reed acts like a comb made of metal stripes. A typical reed is shown in Figure 3.6.

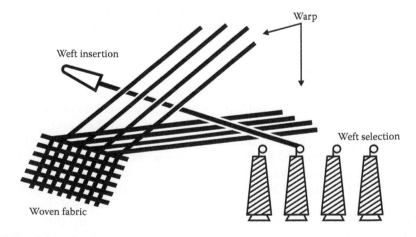

FIGURE 3.5
Picking. (Reproduced from S. Adanur, *Handbook of Weaving*. Lancaster, PA: CRC Press, 2000. With permission.)

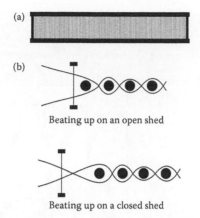

FIGURE 3.6
(a) A reed used for beat-up, (b) beat-up for an open and closed shed. (Reproduced from S. Adanur, *Handbook of Weaving.* Lancaster, PA: CRC Press, 2000. With permission.)

3.2.2 Secondary Motions

The weaving motions required to make the weaving process continuous are called secondary motions. These motions include warp let-off motion and cloth take-up motions.

3.2.2.1 Warp Let-Off Motion

As the fabric is produced, it is required to let off the warp yarn for continuous weaving. The delivery of warp yarn at required speed is called warp let-off motion. The warp yarns are delivered in the form of sheet from weaver beam installed at the back of loom. The let-off motion has been controlled by dead weight called lingos, but nowadays the speed of this motion is controlled using load cell and servo motor. An electrically controlled warp let-off motion is shown in Figure 3.7. As the cloth is woven, the warp yarns exert a tension on whip roller. The whip roller moves forward toward the front side of loom and does an amount of work against the force of spring. The work done in terms of displacement is measured by a sensor, which gives signal to control panel. The control panel sends instruction to servo motor to adjust the speed in order to let off the warp sheet.

3.2.2.2 Cloth Take-Up

The woven cloth needs to be wound on a specific package after it has been beaten up. The winding of woven cloth is called take-up. The cloth is wound on a roller, which is placed on the front side of loom, called the take-up roller. The take-up motion defines the pick density of woven cloth. It is important to note here that take-up of cloth is always less than the length of warp sheet

FIGURE 3.7
Let-off mechanism on loom.

due to warp shrinkage. Modern cloth take-up systems are electrically controlled by servomotor as shown in Figure 3.8. The take-up roller is connected to servo motors via pairs of worm and worm wheel. The take-up system is equipped with electrical sensor to control the surface speed of take-up roller to provide the required number of picks per unit length.

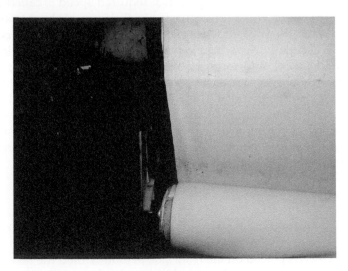

FIGURE 3.8
Take-up mechanism on loom.

3.2.3 Auxiliary Motions

These mechanisms are useful to produce defect-free woven fabric production. Weaving machine is the complex machine. It is difficult to monitor all the points like yarn breaks, finish of weft yarn, etc. Without these tertiary motions, the process will continue, but it is quite impossible to make a defect-free cloth. Hundreds of yarns are running in a loom, so it is quite impossible to monitor all the yarns separately. It may cause the faulty production.

3.2.3.1 Warp Stop Motion

Warp stop motion stops the loom at the event of warp yarn breakage. The motion helps to remove the faults which are expected to be produced due to warp yarn breakage. All the warp yarns are required to pass through an individual special inclined shape wire, which is called dropper. The length of dropper ranges from 120 to 180 mm, while the width of dropper is usually found as 11 mm. In the event of warp breakage, the dropper wire falls on dropper rod. The dropper rod is composed of positive and negative terminals. After the falling of dropper wire, the electrical circuit of the dropper rod is completed. The completion of electrical circuit sends the instruction to servo motor to stop via control panel.

3.2.3.2 Weft Stop Motion

Weft stop motion has been used to stop the loom at the event of weft breakage. In modern looms, mainly two types of weft stop motions are used, namely piezoelectric electronic weft stop sensor and optical sensors. The optical type of weft stop sensors is shown in Figure 3.9. The piezoelectronic weft stop sensor is designed for rapier and projectile looms, whereas the optical sensors are made especially for air-jet looms. The piezoelectronic sensor is made of smart materials, which works on the principle that vibration produces electric charges. The electric charges produced are used to send the signal to stop loom. Under normal running of loom, the electric charges are produced with low amplitude due to less vibration; however, when the weft yarn is broken, a jerk is produced which results in high amplitude of electric charges. These high-amplitude electric charges are used to stop the loom. On the other hand, the optical sensor detects the light emitted by a light source. In air-jet looms, optical weft stop motion sensor serves two purposes, that is, stops the loom if weft yarn is broken and stops the loom if weft yarn has been moved too forward. The sensors are classified as Weft Feeler 1 and Weft Feeler 2. Weft Feeler 1 senses the absence of weft yarn and stops the loom, whereas Weft Feeler 2 senses the presence of yarn and stops the loom.

Other auxiliary motions are warp tension compensation motion, weft tension control motion, auto pick finding motion, weft mixing motion, weft holding, tucking and trimming motion, warp protector motion, weft replenishment motion, and temple motion.

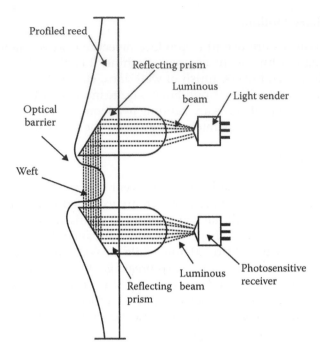

FIGURE 3.9
Optical weft stop sensor. (Textile Learner. Reprinted with permission.)

3.3 Weaving Loom Classification

The weaving loom can be classified based on shed opening system or weft insertion system, which are explained in the following section.

3.3.1 With Respect to Picking Mechanism

Weaving loom classifications with respect to picking systems are shown in Figure 3.10.

The weaving looms are mainly classified into two distinct categories depending on the type of picking or weft insertion system. These categories include shuttle loom and shuttleless loom. The shuttle loom carries a shuttle to insert the weft. The shuttle is comprised of wooden body having a weight of 150–250 grams and a pirn required to wind the weft yarn on it. The pirn is placed inside the wooden body of shuttle. The limitations of shuttle loom are its low speed, heavy weight, and extra stoppage due to empty pirn replacement. The shuttle looms are being rapidly replaced by shuttleless looms. The shuttleless looms are further classified

Weaving loom classifications	Single phase	Shuttle	Hand loom
			Power loom
			Automatic loom
		Shuttleless	Projectile
			Rapier
			Air-jet
			Water-jet
Multi phase	Warp wave		
	Weft wave		

FIGURE 3.10
Weaving loom classifications with respect to picking systems.

into distinct categories including projectile loom, rapier loom, air-jet loom, and water-jet loom. All the possible weft insertion systems are shown in Figure 3.11.

3.3.2 With Respect to Shedding Mechanism

The weaving looms are classified into three distinct categories depending on shedding mechanism: tappet loom, dobby loom [3], and jacquard loom. A summary of weaving loom with respect to shedding mechanism is given in Table 3.1. In tappet loom, the healed frames are operated with the help of tappet cams, which are designed according to weave designs. The advantages of tappet loom are simple mechanism, low initial cost, and easy maintenance. The number of cams, in tappet loom, depends on the weave repeat, for example, 4 cams for 3/1 twill. The limitations of tappet loom are; tappet cam designed for one weave cannot be used for other weave and long time is required for cam change. In dobby looms, the healed frames are operated by jacks and levers, and lifting and lowering of healed frames are controlled by pattern chain. The dobby looms can be equipped with a maximum of 32 healed frames, which expands the weave design possibilities. The weave designs such as crepe, honey comb, huckaback, mockleno, and bedford cord are possible with dobby looms. However, mechanism is complicated as compared to tappet loom, and initial cost and maintenance cost are high. In jacquard loom, the warp yarns are individually controlled by a cord called harness cord rather than healed frame. The jacquard loom further expands the weave design possibilities to an unlimited extent. The woven cloth with complicated designs such as animals, flowers, and geometrical figures can be produced. A summary of available jacquard systems from Staubli is given in Table 3.2.

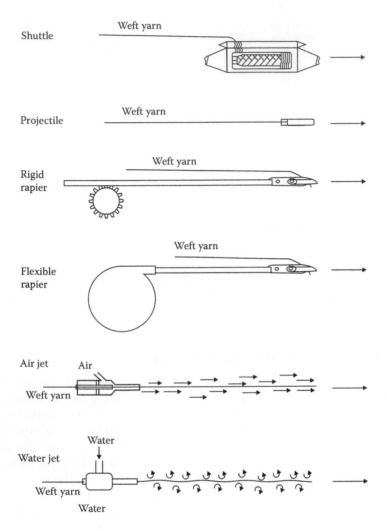

FIGURE 3.11
Weaving loom weft insertion media. (Reproduced from S. Adanur, *Handbook of Weaving*. Lancaster, PA: CRC Press, 2000. With permission.)

3.4 Woven Fabric Geometrical/Graphical Representation

3.4.1 Nomenclature

It has always been a challenge to artists to convert their imaginary thoughts into a representable form, but science always simplifies this, where their thoughts can be easily represented on a paper. Thus, the woven design

TABLE 3.1

Summary of Weaving Looms with Respect to Shedding Mechanism

Sr. No.	Type of Loom	Repeat Length (max.)	Number of Harness (max.)	Pitch (mm)
1	Tappet (positive and negative)	Up to 8 picks	12	18
2	Dobby (positive mechanical)	6000	28	12
3	Dobby (negative mechanical)	150	16	12
4	Dobby (rotary mechanical)	4700	28	18
5	Dobby (negative electronic)	6400	16	12
6	Dobby (positive electronic)	6400	28	12
		Number of Hooks		
7	Jacquard (single lift mechanical)	600		
8	Jacquard (double lift mechanical)	1200		
9	Jacquard (double lift electronic)	Up to 14,000		

can also be represented on a paper, which is helpful to produce fabric on industrial scale. Normally, the warp yarns are represented by vertical lines, whereas the horizontal lines represent the weft yarns. The up position of warp over weft yarns is indicated by a cross (x), whereas the down position of warp under weft yarns is indicated by dot (.) or leave it as blank. The graphical representation of plain weave is shown in Figure 3.12a. The cross-sectional view of plain weave is shown in Figure 3.12b.

3.5 Elements of Woven Fabric Structure

3.5.1 Weave

The weave is a technical design of woven fabric which defines how warp and weft yarns are interlaced with each other. There are three main basic weaves, namely plain, twill, and satin, as shown in Figure 3.13.

3.5.2 Draft

The draft is a technical representation of weave design used to indicate which warp yarn goes to which healed frame. The draft of 3/1 twill is shown in Figure 3.14. Since the given weave design contains four distinct warp yarns, each warp yarn will go to a separate healed frame. However, the same warp yarn can be inserted to the same healed frame.

TABLE 3.2

Summary of Staubli Jacquard

Sr. No.	Application	Recommended Jacquard (Staubli)	Number of Hooks	Characteristics
1	Ribbons and labels	CX 182/LX 32/ LX 62	192/448/896	Needle weaving, narrow fabrics, double-lift open-shed jacquard machine with crank drive, inclined shed on LX range
2	Written selvedges	CX 172	64 or 80	Labeling and logos on selvedges, double-lift open-shed jacquard
3	Fabric inscriptions on selvage	CX 182	96 or 192	machine, crank shed movement, parallel shed
4	Flat fabrics and terry cloth	LX 1602 and LX 3202	LX 1602: 3072, 4096, 5120 hooks. LX 3202: 6144, 8192, 10240, 12288(x2), 14336, 18432 hooks	Flat fabrics, terry cloth, and technical fabrics, double-lift open-shed jacquard machine, Coaxial shaft system, controlled by complementary cams, Fine adjustment of shed opening and shed angle
		SX	1408 hooks (11 rows of 16 modules) 2688 hooks (21 rows of 16 modules)	Flat fabrics and terry cloth, 76 precise, accessible shedding settings
5	Labels and name selvedges on fabric	UNIVALETTE	64 or 96 actuators per head	Labeling and logos on selvedges and in all flat fabrics, the jacquard machine has its own drive with no mechanical link to the weaving machine and is electronically synchronized
6	Free of mechanical constraints	UNIVAL 100	512–15360(x2) threads with all intermediate formats	Flat fabrics, terry cloth, technical fabrics and carbon fiber fabrics, individual warp thread control with Stäubli JACTUATOR, the jacquard machine has its own drive with no mechanical link to the weaving machine and is electronically synchronized
7	Velvet fabrics	SX V	2688 hooks	Velvet fabrics, two parallel shafts, cam driven on both sides
		LX 1692, LX 3292	3072, 4096, 5120 hooks 6144, 8192, 10240, 12288, 14336 hooks	Coaxial shaft system, controlled by complementary cams Fine adjustment of shed opening and shed angle

(a)   (b)

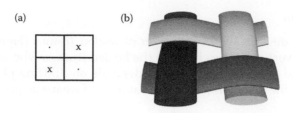

FIGURE 3.12
(a) Cross and dot representation of up and down position (respectively) of warp yarn in plain weave, (b) cross section view of plain weave.

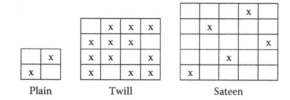

Plain   Twill   Sateen

FIGURE 3.13
Basic weaves.

Repeat size = 12 ends × 8 picks
No. of frames = 4

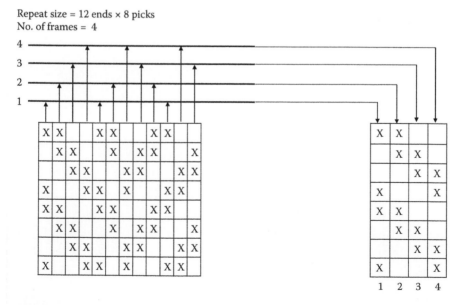

FIGURE 3.14
Design, draft, and peg plan of a given weave.

3.5.3 Peg Plan

The peg plan denotes the lifting order of healed frames (Figure 3.14). In a peg plan, the vertical spaces indicate the healed frames, whereas horizontal spaces indicate the pick. The peg plan depends on drafting plan. In case of straight draft, the peg plan will be the same as of weave deign.

References

1. S. Adanur, *Handbook of Weaving*. Lancaster, PA: CRC Press, 2000.
2. M. Mahfuz. (March 15). *Different Types of Shedding Mechanism in Weaving Process*. Available online at: http://textilelearner.blogspot.com/2013/09/different-types-of-shedding-mechanism_4.html.
3. International-Textbook. (2002, March 15). *Dobbies*. Available online at: https://www.cs.arizona.edu/patterns/weaving/monographs/ics495.pdf.

4

Conventional Woven Structures

Muhammad Umar Nazir and Yasir Nawab

CONTENTS

4.1 Basic Weaves

Weave is the interlacing pattern warp and weft yarns, in order to produce a woven fabric. These patterns are set according to the required properties of the finished fabric. The wide spectrum of these patterns ranges from simple to complex interlacements, which imparts specific functions of the finished fabric. In this section, basic weaves will be discussed. The basic weaves are plain, twill, and satin. All the others are derivatives of these basic weaves or their combination.

4.2 Plain Weave

Plain is the simplest weave, in which warp and weft threads interlace in alternate manner (as shown in Figure 4.1), giving maximum number of interlacements. This maximum interlacement imparts firmness and stability to the structure. In trade, the special names like broadcloth, taffeta, shantung, poplin, calico, tabby, and alpaca are applied to plain weave. At least two ends and two picks are required to weave its basic unit. A minimum of two heald frames are required for this weave, but more than two (multiple of basic weave) heald frames can be used to weave this construction.

4.2.1 Warp Rib

Warp ribs are a modified form of plain weave. It has 1/1 interlacements in the filling direction, which differs from the simple plain weaves. This modified

FIGURE 4.1
Plain weave (1/1).

interlacement results in the formation of cords, ridges, or texture across the warp direction of the fabric. These cords or ridges are formed due to the grouping of the filling yarns. The repeat of warp rib is always on two warp yarns. The first warp yarn follows the formula, while the second warp yarn is in the opposite direction of the first one. It requires two heald frames at least, but multiple of these can also be employed. The number of weft yarns in a repeat unit of this weave is equal to the sum of the digits in formula of warp rib. For example, 2/2 warp rib requires 2 warp yarns and 4 weft yarns. Design of the above-stated warp rib is shown in Figure 4.2. Warp rib is also known as ottoman.

4.2.1.1 Regular Warp Rib

This extension of warp rib is made when the numerator and denominator of weave formula have the same value as stated above in the example of 2/2. The other examples of regular warp ribs are 3/3 and 4/4.

4.2.1.2 Irregular Warp Rib

If the numerator and denominator of weave formula of warp rib have different values, then the resulting design will be irregular warp rib. For example, 2/3 warp rib is an irregular warp rib, and its design is shown in Figure 4.3.

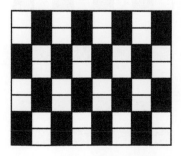

FIGURE 4.2
Warp rib (2/2).

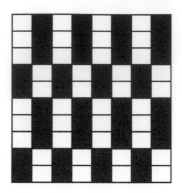

FIGURE 4.3
Irregular warp rib (2/3).

If only a portion of the formula has the same value of numerator and denominator, the design is still considered to be an irregular warp rib. For example, 1/2-1/1-3/2 is an irregular warp rib. Its design is shown in Figure 4.4.

4.2.2 Weft Rib

Weft ribs are another modified form of plain weaves. It has 1/1 interlacements in the warp direction, which differs from the simple plain weaves. This modified interlacement results in the formation of cords, ridges, or texture across the weft direction of the fabric. These cords or ridges are formed due to the grouping of the warp yarns. The repeat of weft rib is always on two weft yarns. The first weft yarn follows the formula, while the second weft yarn is in the opposite direction of the first one. It requires two heald frames at least, but multiple of these can also be employed. The number of warp yarns in a repeat unit of this weave is equal to the sum of the digits in formula of warp rib. For example, 2/2 weft rib requires 2 weft yarns and 4

FIGURE 4.4
Irregular warp rib (1/2+1/1+3/2).

FIGURE 4.5
Weft rib (2/2).

warp yarns. Design of the above-stated warp rib is shown in Figure 4.5. Weft rib is also known as half panama.

4.2.2.1 Regular Weft Rib

This extension of weft rib is made when the numerator and denominator of weave formula have the same value as stated in the above example of 2/2. The other examples of regular weft ribs are 3/3 and 4/4.

4.2.2.2 Irregular Weft Rib

If the numerator and denominator of weave formula of weft rib have different values, then the resulting design will be irregular weft rib. For example, 2/3 weft rib is an irregular weft rib. Design of the said weave is shown in Figure 4.6.

If only a portion of the formula has the same value of numerator and denominator, the design is still considered to be an irregular weft rib. For example, 1/3-2/1-3/1 is an irregular weft rib. Its design is shown in Figure 4.7.

4.2.3 Matt Weave

This type of weave is constructed by extending the plain weave in warp and weft directions at the same time so that two or more threads work alike in

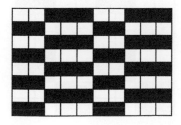

FIGURE 4.6
Irregular weft rib (2/3).

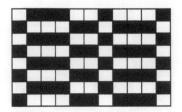

FIGURE 4.7
Irregular weft rib (1/3+2/1+3/1).

both directions. In this weave, the same size of squares appear on both sides of the fabric showing the same number of warp and weft yarns on front and back of the fabric. Matt weave is also commercially known as basket, hopsack, or full panama. This weave requires a minimum of two heald frames. Design of the 2/2 matt weave is shown in Figure 4.8. The matt weaves can be extended further to give more prominence but restricted due to loose structure and modified in several ways. In matt weave, the warp ends that work alike tends to twist around each other. To avoid this twisting of the yarns, warp ends that work alike are drawn from different slits of the reed.

4.2.3.1 Regular Matt Weave

The matt weave that shows equal spaces on both sides of the fabric represents regular matt weave. This type of interlacement refers to loose constructions so that employed only in fine fabrics. The above design of the 2/2 matt weave is regular.

4.2.3.2 Irregular Matt Weave

The matt weave that shows unequal spaces on both sides of the fabric represents irregular matt weave. The irregular matt weave design of 2/3 is shown in Figure 4.9.

FIGURE 4.8
Matt weave (2/2).

FIGURE 4.9
Irregular matt weave (2/3).

FIGURE 4.10
Fancy matt weave (3/1+1/3).

4.2.3.3 Fancy Matt Weave

The matt weave has loose structure due to less interlacement, so sometimes it is combined with warp or weft ribs. This modified matt weave is known as fancy matt weave. A fancy matt weave of 3/1-1/3 is shown in Figure 4.10.

4.3 Twill Weave

Twill weave is another basic weave which is well known for its diagonal line formation in the fabric due to its interlacing pattern. This weave and its derivatives are used for the ornamental purposes. Twill has closer setting of yarns due to less interlacement imparting greater weight and good drape as compared to the plain weave. In simple twill, the outward and upward movement of the interlacing pattern is always one that imparts a diagonal line to this design. The direction of the propagation of twill line classifies twill into right-hand or left-hand twill.

FIGURE 4.11
Right hand twill (3/1).

In right-hand twill (Figure 4.11), the diagonal line runs from the lower left corner to the upper right corner of the design. This direction of the twill is also referred to as Z direction. If the twill lines propagate from the lower right to the upper left corner of the design, it is called left-hand or S-direction twill (Figure 4.12). The warp float coincides to the weft float and weft float to the warp float on opposite sides of the fabric, which means that if warp is prominent on one side of the fabric, then weft will be prominent on the other side of the fabric and vice versa. For example, 3/1 is warp faced (Figure 4.13) having back of the fabric opposite, that is, 1/3 or weft faced (Figure 4.14).

Twill weaves which have to show more prominent warp are constructed with a greater number of warp yarns as compared to weft yarns per unit space. Similarly, if twill is weft faced, it will have a more number of weft yarns as compared to warp yarns in a unit space. Twill weaves are also classified into regular or balanced and irregular or unbalanced weaves. The digits in the regular or balanced twill will be equal and vice versa. For example, 2/2, 3/3, and 4/4 are regular or balanced twill (Figure 4.15), whereas 2/1, 3/1, and 1/2 are irregular or unbalanced twills (Figure 4.16). Twill weave requires at least three heald frames (e.g., 2/1 or 1/2) to produce the fabric.

The twill diagonal line makes an angle that depends on the thread density of warp and weft. This twill angle varies from 15° to 75°. If the twill makes

FIGURE 4.12
Left hand twill (3/1).

FIGURE 4.13
Warp faced twill (3/1).

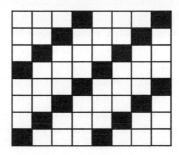

FIGURE 4.14
Weft faced twill (1/3).

FIGURE 4.15
Balanced twill (2/2).

an angle of exactly 45°, it is known as common twill. The twill will be called steep twill if the diagonal line makes an angle greater than 45°, while it will be reclining twill if the angle of twill line reduces from 45°.

To get the repeat of the twill, add all the interlacing points above the line (warp up) to the sum of interlacing points below the line (weft up). For example, 3/1 twill will have a repeat on four ends and four picks. The term drill is associated with 3/1 twill weave.

FIGURE 4.16
Unbalanced twill (3/1).

Plain weave has more interlacements as compared to twill weave, so the contact surface of the threads in plain weave is more due to excessive interlacements. This provides plain weave more tensile resistance as compared to the twill weave. Twill weaves have longer floats as compared to the plain, so the slippage of threads is more in this weave when subjected to tear. So, twill weave has more tear resistance relative to plain weave. More the interlacements, more will be the firmness of the structure along with stiffer appearance. Plain weave has poor drape and luster relative to the twill weave.

Twist is required to spin the fibers into a yarn. This twist imparts specific features to the final yarn, which in turn affect the fabric that are made from these yarns. More twisted yarn is used in warp as compared to weft. The more the twist, the more will be the strength (optimum level) and hardness of yarn but lesser is the luster. The twist of yarn also affects the prominence of the twill made from these yarns. If the direction of twist and the direction of twill are the same, then the forming twill will have lower prominence. To impart more prominence to the twill, the direction of the twill and the direction of the twist must be opposite to each other. In case of a herringbone twill, where twill has both right and left directions, the prominence in one direction will be more than the other.

4.3.1 Pointed Twill

This is the most commonly used derivative of twill weave. It is formed by reversing the twill line at specific intervals to combine left-hand and right-hand twills in a single design. It is also known as zigzag or waved twill. The reversing of the twill may be done horizontally or vertically.

In pointed twill, first of all, draw the basic weave. The last end of the basic weave is known as pivot point where design reverses. While reversing the design at this point, leave the pivot end and draw the remaining ends as it is in the sequence. Follow this process until the design repeats itself. This is the process to make the horizontal pointed twill. The same procedure will be done to draw vertical pointed twill. The procedure is illustrated with the design in Figure 4.17.

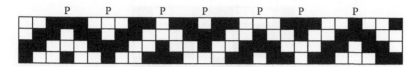

FIGURE 4.17
Pointed twill, 2/2 (4R, 3L).

Pointed twill is also divided into regular and irregular. In regular pointed twill, the amount of movement in left and right directions is the same, whereas in irregular pointed twill, the movement of the twill is more in left or right direction as compared to the others. In the case of regular pointed twill, the repeat of the design will be equal to the sum of left-direction movement and right-direction movement. For example, in the case of 2/2 regular pointed twill with 4R and 4L (equal movement on both sides), the design will repeat on 8 ends and 4 picks (horizontal pointed twill). In the case of irregular pointed twill of 2/2 with 4R and 3L (unequal movement on both sides), the design will repeat on 28 ends and 4 picks (horizontal pointed twill).

4.3.1.1 Horizontal Pointed Twill

This type of twill is formed when reversing of the diagonal twill line occurs on warp ends. In this case, the number of ends is more than the picks. In this type of twill, pointed draft is used. The example of the horizontal pointed twill shown in Figure 4.18 is 2/2 basic twill with 4R and 4L.

4.3.1.2 Vertical Pointed Twill

The vertical pointed twill is formed when reversing of the twill line is done on the picks. The number of picks will be more than the warp ends in this case in a unit space. This type of design sometimes requires dobby shedding due to excessive picks which cannot be handled on simple tappet looms. An example of vertical pointed twill is shown in Figure 4.19, which is 2/2 vertical pointed twill with 4R and 4L.

FIGURE 4.18
Horizontal pointed twill, 2/2 (4R, 4L).

FIGURE 4.19
Vertical pointed twill, 2/2 (4R, 4L).

4.3.2 Herringbone Twill

This twill also depends on the reversal of the direction to achieve the distinct effect. Their method of construction is different from the waved twills. First draw the simple basic weave, and then at the pivot point draw the opposite of the pivot point, and carry on the twill in the opposite direction of the first twill direction. Due to this opposition, the warp-faced twill in one portion changed to weft-faced twill in the second portion from the pivot point or the reversal point. A herringbone twill of 3/1 with 4R and 3L is shown in Figure 4.20.

Herringbone twill is also divided into regular and irregular twill. In regular herringbone twill, the amount of movement in left and right directions is the same, whereas in irregular herringbone twill, the movement of the twill is more in left or right direction as compared to the others. In the case of regular herringbone twill, the repeat of the design will be equal to the sum of left-direction movement and right-direction movement. For example, in the case of 2/2 regular herringbone (Figure 4.21) twill with 4R and 4L (equal movement on both sides), the design will repeat on 8 ends

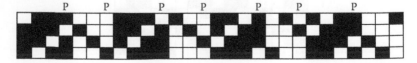

FIGURE 4.20
Herringbone twill, 3/1 (4R, 3L).

and 4 picks (horizontal herringbone twill). In the case of irregular herringbone (Figure 4.22) twill of 2/2 with 4R and 3L (unequal movement on both sides), the design will repeat on 28 ends and 4 picks (horizontal herringbone twill).

4.3.2.1 Horizontal Herringbone Twill

Herringbone twill extended in warp direction is known as horizontal herringbone twill. This twill is used most commonly in shirts and overcoats. A horizontal herringbone twill of 3/3 is shown in Figure 4.23 with 4R and 4L.

4.3.2.2 Vertical Herringbone Twill

Basic twill extended in weft direction is known as vertical herringbone. This twill is employed in fabrics for hanging and in soft furnishing. A vertical herringbone twill of 3/3 is shown in Figure 4.24 with 4R and 4L.

4.3.3 Skip Twill

This derivative of twill forms the twill lines in the same direction. The process of drawing skip twill is much similar to the herringbone twill.

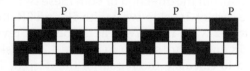

FIGURE 4.21
Regular herringbone twill, 2/2 (4R, 4L).

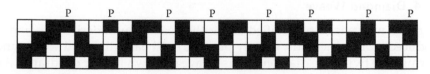

FIGURE 4.22
Irregular herringbone twill, 2/2 (4R, 3L).

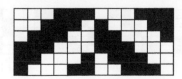

FIGURE 4.23
Horizontal herringbone twill, 3/3 (6R, 6L).

FIGURE 4.24
Vertical herringbone twill, 3/3 (6R, 6L).

The only difference of the skip from the herringbone is the twill line progression in the same direction at the reversal. The warp-faced twill will convert to weft face at reversal, but the direction of both these twills will be the same. For example, in the case of 2/3 skip twill with 5R and 3R, the weft-faced 2/3 twill will change to 3/2 warp-faced twill at the point of reversal. Designs of the 2/3 skip twill 5R and 3R are shown in Figure 4.25. In skip twill, there is no specific formula to calculate the repeat of the design. To check the full repeat, the design has to carry until its repeat occurs.

4.3.4 Diamond Weave

Diamond weave is also classified under the head of twill derivatives and is produced by two methods, namely pointed twill base diamond weave and herringbone twill base diamond. This weave is drawn in four quadrants. All the quadrants jointly form a repeat unit of a diamond weave. The name is given to this weave due to the formation of unique diamond-like shape.

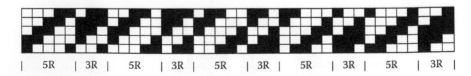

| 5R | 3R | 5R | 3R | 5R | 3R | 5R | 3R | 5R | 3R |

FIGURE 4.25
Skip twill, 2/3 (5R, 3R).

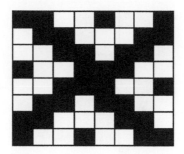

FIGURE 4.26
Pointed twill base diamond weave, 2/2.

4.3.4.1 Pointed Twill Base Diamond

It is based on pointed twill. This weave is symmetrical in its vertical and horizontal axes. The repeat of the design is calculated by adding the digits of the basic weave and then multiplying them by two. The answer will be the number of ends and picks in the final design. For example, in 2/2 pointed twill base diamond weave, the repeat ends will be 8 and repeat picks also the same. First of all, draw the basic weave in first quadrant, and then take the last end of basic weave as pivot point and draw pointed twill in second quadrant. Repeat the same procedure in third and fourth quadrants. The final designs are shown in Figures 4.26 and 4.27.

4.3.4.2 Herringbone Twill Base Diamond

This is the diamond weave that uses the herringbone in its formation. This weave is symmetrical in its diagonal axis. Diagonally opposite quadrants are similar to each other. The repeat calculation of the design is the same as that in pointed base diamond weave. It also requires four quadrants to complete design. The first quadrant is for the basic weave, and then in the second quadrant, the herringbone twill is drawn considering the last end of the basic weave as a reversal or pivot point. The same procedure will be adopted for the rest

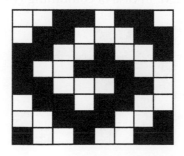

FIGURE 4.27
Pointed twill base diamond weave, 2/2.

FIGURE 4.28
Herringbone twill base diamond weave, 2/2.

of the quadrants to complete the design. A diamond weave design based on 2/2 herringbone is shown in Figure 4.28. By using balanced twill like 2/2, 3/3, and so on, a well-balanced diaper effect can be produced. Due to this reason, diamond weaves based on herringbone twill are also called diaper weaves.

4.3.5 Combination Twill

In this twill type, a number of weaves are grouped in a single end. This will give the combined effect of these weaves. The repeat of the combination twill will be equal to the aggregate of constituting weaves. For example, the repeat of 2/1–2/2 will be on 7 ends and 7 picks. The design of a combination weave is shown in Figure 4.29.

4.3.6 Combined Twill

Combined twill is different from the combination twill. In this twill, the different twills are constructed on alternate ends (method 1) or alternate picks (method 2). In this type, least common multiple (LCM) of the different weaves is taken. All the different weaves are drawn on these ends and picks separately. The design repeat will double the LCM in the direction in which design is constructed either end wise or pick wise. The other side of design repeat is equal to the LCM. In case of combining two twill weaves of 4 and

FIGURE 4.29
Combination twill, 2/1+1/2.

FIGURE 4.30
Combined twill, 3/1+3/2.

5 threads, LCM will be 20. Now, if it is drawn in warp direction, then both twills will be constructed on 20 ends each with a total of 40 ends. The pick repeat will be on 20 picks. If the design is constructed in pick direction, then the pick repeat will be on 20 picks each with a total of 40 picks. The end-wise repeat of design in this case will be 20, which is the LCM of the 2 twills. Both examples are shown in Figures 4.30 and 4.31.

4.3.7 Broken Twill

This twill is achieved by breaking the regular twill at specific intervals. A variety of distinctive designs can be drawn by using this weave. The regular twill can be broken in a number of ways, but the most simple and common way is to stop the regular twill and reverse the next ends according to the break unit. In case of drawing six ends broken twill with a break unit of three, its order will be straight three ends 1, 2, and 3 and after that last three ends starting from the last end like 6, 5, and 4. Using well-balanced twill, the broken effect can be achieved in balanced effects. The repeat of the broken twill is the same if the break unit is a factor of the original weave; otherwise, calculate the LCM of basic weave and the break unit to get the repeat of the broken twill design. For example, if original weave is on eight ends and break unit is two, then the repeat of the broken twill will be eight. If break unit is 3 in this case, then the LCM that is 24 will be the repeat of the broken twill. To draw broken twill, follow the instructions given below.

Run the basic twill in one direction up to the break unit.

Add break unit in last end of the above-mentioned direction and draw reverse direction twill from that end. Follow this procedure until you get the starting point again. This procedure is illustrated in Figures 4.32 and 4.33.

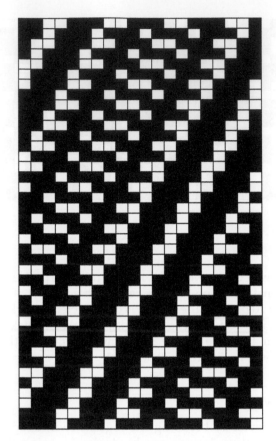

FIGURE 4.31
Combined twill, 3/1+3/2.

4.3.8 Elongated Twill

The rate of advancement in regular twill is normally one step ahead and one step forward. In elongated twills, the rate of advancement of the twill is changed. The change may be in both directions as per requirement. The advancement will be two at least or more according to the requirement.

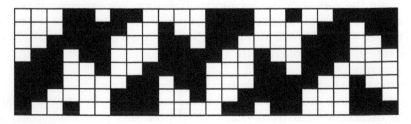

FIGURE 4.32
Broken twill, 4/4.

FIGURE 4.33
Broken twill, 4/3+3/2.

Elongated twill can be drawn if the rate of advancement is known or by rearranging the ordinary twill in certain order. The number counted in each case is a factor of the number of threads in the given twill.

Method 1: When the rate of advancement is known. The designs which are 2/2+1/3 and 2/2 with advancement of 2 warp wise and weft wise, respectively, are shown in Figures 4.34 and 4.35.

Method 2: When a basic weave is given. Rearrange this weave as per the requirement. Both warp- and weft-wise elongated weaves can be drawn as per requirement. The designs are the rearrangements of basic weave of 7/5 in every second and third ends (warp wise) and second and third picks (weft wise), respectively. These designs are shown in Figures 4.36 through 4.39.

4.3.9 Transposed Twill

Transposed twill is also known as rearranged twill. In this twill, only rearrangement of the original twill weave is done to get more attractive designs.

FIGURE 4.34
Elongated twill, 2/2+1/3.

FIGURE 4.35
Elongated twill, 2/2.

FIGURE 4.36
Elongated twill, 7/5.

The transposition just interrupts the basic twill line at specific intervals according to the break unit. The process of transposition is much similar to the breaking of the twill. The weave repeats on the same number of ends as the original weave if break unit is a factor of the original weave. The weave repeat will be the LCM of the break unit and the original weave if the break unit is not a factor of the original weave. The process is illustrated by examples given in Figures 4.40 through 4.42. In all of the examples, basic weave is 4/4 twill having break units 3, 4, and 2.

4.4 Satin/Sateen

Satin/sateen is a basic weave that does not have any regular pattern like twill. The surface of the fabric is either warp or weft faced. Satin is warp faced,

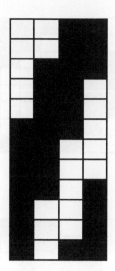

FIGURE 4.37
Elongated twill, 7/5.

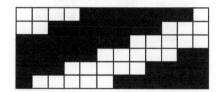

FIGURE 4.38
Elongated twill, 7/5.

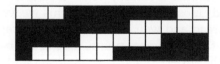

FIGURE 4.39
Elongated twill, 7/5.

FIGURE 4.40
Transposed twill, 4/4.

FIGURE 4.41
Transposed twill, 4/4.

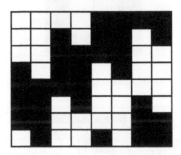

FIGURE 4.42
Transposed twill, 4/4.

which means that all the surface of the fabric will show the warp threads except for the one thread interlacement with other series of yarn. If it is weft faced, then it will be known as sateen, which means that fabric surface will show the weft threads mostly. The unique in this weave is the single interlacement of warp thread and weft thread in a single repeating unit. These weaves have the least interlacement points among the basic weaves. Due to this reason, it gives the surface of fabric more luster and smoothness. Along with these properties, more close packing of the threads is possible, which gives the maximum achievable cover factor in this weave. There is a specific procedure of making this type of weaves. It will follow a move number to draw a design. The number of threads is given to construct a weave. From that number of threads in a weave, select specific move number to construct the weave. To select the move number, follow the instructions listed below:

First write the numbering up to the required number of ends, that is, 1, 2, 3, 4, and 5.

Leave number 1 because it will make twill.

Leave the second last and last numbers.

Leave all the factors of the weave.

Leave all the numbers exactly half of the weave.

Leave the numbers that have common divisible with the weave.

Following these rules satin or sateen weaves can be constructed with ease. Examples of the weaves are shown in Figures 4.43 through 4.45.

The satin or sateen weaves that do not follow any specific move number are termed as irregular satin and sateen weaves. Four and six threads satin and sateen do not follow any specific move number so termed as irregular. These are the only irregular satin or sateen weaves. To construct these weaves, combination of move numbers is used. Design of these weaves is shown in Figures 4.46 and 4.47.

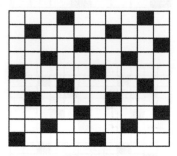

FIGURE 4.43
Sateen, 5 end.

FIGURE 4.44
Satin, 5 end.

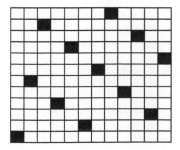

FIGURE 4.45
Sateen, 11 end.

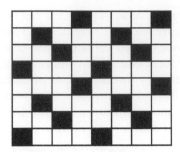

FIGURE 4.46
Irregular sateen, 4 end.

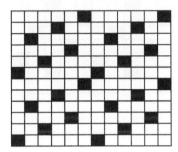

FIGURE 4.47
Irregular sateen, 6 end.

4.5 Honey Comb Weave

This name is given to this weave due to its honey bee web-like structure. It makes ridges and hollow structures which finally give a cell-like appearance. In this weave, both warp and weft threads move freely on both sides, which coupled with rough structure. The fabric made by this weave has longer float all over the fabric. Due to this reason, it is radially absorbent of moisture. This property made these weaves useful for towels, bed covers, and quilts. This weave is further divided into three types which are explained below. Most commonly, these weaves are constructed on repeats which are multiple of four in ends and picks.

4.5.1 Single-Ridge Honey Comb

It is the most simplest among all the honey combs. In this case, the weave repeat will be equal to $1/x$ where x will be one less than the repeat size. If there is a repeat of (12×12), 12 ends and picks each, then x will be equal to 11 and the weave will be $1/11$. In this weave, first draw right-hand twill and then draw left-hand twill in such a way that clear cross is made. Then, put dots on opposite sections of the twill crossing lines. Draw x in the remaining

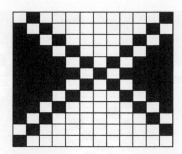

FIGURE 4.48
Single ridge honey comb, 12 × 12.

boxes of the honey comb. Clear cross is formed when *x* of the right-hand twill coincides with *x* of the left-hand twill. Clear cross is made by starting the left-hand twill from the second pick or from the second last end. An example of the single ridge 12 × 12 is shown in Figure 4.48.

4.5.2 Double-Ridge Honey Comb

In the case of double-ridge honey comb, the weave will be 1/1.1/*x* where *x* is three less than the repeat. So, drawing a weave on 16 × 16, the weave will be 1/1.1/13. The method of drawing is the same as that in single-ridge honey comb. An example of this design is shown in Figure 4.49.

4.5.3 Brighton Honey Comb

Formations of Brighton of honey comb follow the rules listed below:

1. Make Z twill of 1/*x* where *x*= Repeat – 1.
2. Make S twill of 1/1. 1/*x*, where *x* = Repeat – 3.
3. Keep in mind to form clear cross while making S twill.

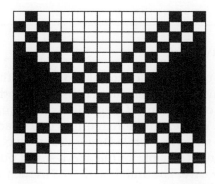

FIGURE 4.49
Double ridge honey comb, 16 × 16.

FIGURE 4.50
Brighton honey comb, 24 × 24.

4. Select the longest float between two crosses up and below the Z twill line, that is, if on one side of Z twill, one cross is below, then on the other side of the twill, that cross must be below.

5. Take these selected crosses as reference points.

6. Fill reference point line (horizontally) by crosses obtained by the following formula: (Repeat/2) – 1.

7. Now above and below the reference point line, fill with crosses the remaining odd numbers till one.

The application of these rules is illustrated in Figure 4.50.

4.6 Huck a Back Weave

This weave is largely used for cotton towel and linen cloth. It has longer floats in two quadrants, which make them more moisture absorbent so employed in towels. This weave is combination of longer floats of symmetric weaves in two quadrants and plain weaves in the remaining two quadrants. Plain weave gives firmness to the structure, while longer float weave increases the absorbency of fabric, making it suitable for the above-stated purpose. Special draft is employed for this weave. The draft is arranged in such a way that odd ends are drawn in two front heald frames and the even threads are drawn from back two heald frames. The purpose of this special draft is to weave plain

FIGURE 4.51
Huck a back weave.

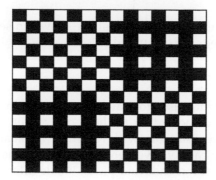

FIGURE 4.52
Huck a back weave.

fabric without redrawing of beam. For this purpose, heald frame one and two are coupled together, and heald frames three and four are coupled together. Sometimes, longer float symmetric weaves are used in combination of plain weaves in huck a back weave, which is also termed as honey comb huck a back weaves. Examples of this weave are shown in Figures 4.51 and 4.52.

4.7 Crepe Weave

Crepe weave refers to those weaves that do not have any specific pattern. These weaves may contain a little bit appearance of twills, but they do not have the prominence. They make small patterns or minute spots and seed-like appearance all over the fabric surface. These weaves may be used separately or in combination with other weaves. Crepe weaves are frequently employed in making the ground of the figured fabrics. In simple words, crepe weave is used to make a rough appearance. If we make crepe weaves with crepe yarns, this combination will give more remarkably pebbly or

FIGURE 4.53
Crepe (sateen based), 7 end.

puckered appearance. Crepe weaves can be drawn in several ways, but the most common methods are discussed below.

4.7.1 Sateen Method

This method is frequently used to draw crepe weaves. In this method, simply sateen weave is drawn according to the requirement. Then, this sateen is filled by the insertion of suitable twill. The twill repeat and the sateen repeat must be the same. In case of seven-end sateen, twill that will be used must have a repeat on the seven ends and picks. While filling the twill in sateen, keep in mind that the starting point of both sateen and twill must be the same. Examples of seven-end and nine-end sateens with twill weaves of 3/1+2/1 and 3/2+2/2, respectively, are shown in Figures 4.53 and 4.54.

4.7.2 1/4 Turn Method

This is the most interesting way to draw a crepe weaves and uses balanced twills like 2/2 or 3/3 as the base design. In this case, the repeat of design is two times the number of ends and picks used in the base twill. For example, in 2/2 twill, the repeat of crepe will be equal to eight ends and picks. First of

FIGURE 4.54
Crepe (sateen based), 9 end.

FIGURE 4.55
Crepe (1/4 turn method), 2/2.

all, draw the twill weave on alternate ends and picks. Take this side as "A." Then, rotate the design by 90°, and take this side as "B." Now, start filling the same twill from this side on alternate ends and picks. Then, rotate the design again by 90°. Take this side as "C," and start filling the boxes with twill weave on alternate ends and picks. After filling this side, rotate the design by 90° for the last time, and fill the twill from this side. This side will be taken as "D." With the completion of twill on this side, crepe is completed. To avoid any confusion while filling the twill on these sides, fill then with respective side names like A, B, C, and D. The process of drawing this weave design is illustrated in Figure 4.55.

4.7.3 Reversing Method

It is one of the simplest methods to draw the crepe weave design. In this method, take any irregular weave design for any motif (motif is one of the quadrants if we take design repeat in four quadrants). Then, take reverse of this design in the second quadrant. Then, take reverse of the second quadrant in the third quadrant and its reverse in the fourth quadrant. The repeat of design depends on the motif. If motif is four, then repeat will be on eight ends and picks (motif × 2). Examples of these designs are shown in Figures 4.56 and 4.57.

4.7.4 Super Imposed Method

In this method of crepe weave, combination of two weaves is taken. While selecting two different weaves, it should be noted that one of the weaves must be irregular. If both of the designs are irregular, they will produce more uneven appearance that is required in crepe. The number of threads in the design repeat will be equal to the LCM of both the weaves used in drawing it. In case of four-end and six-end sateens (both four-end and six-end sateens are irregular), repeat of crepe weave will be 12 ends and picks, which is the LCM of these two weaves. This method is well explained in designs shown in Figures 4.58 and 4.59.

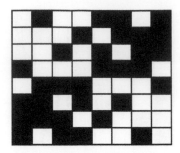

FIGURE 4.56
Crepe (reversing method), motif of 4.

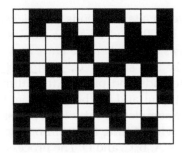

FIGURE 4.57
Crepe (reversing method), motif of 5.

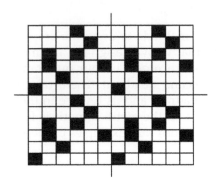

FIGURE 4.58
Crepe (super imposed method), 4 and 6 end sateen.

4.7.5 Plain Method

It is another method to draw a crepe weave. In this method, we select sateen. This sateen is then drawn on alternate ends and picks. It means that 7-end sateen will require 14 ends. Then, fill this sateen with suitable twill weaves of your own choice. The starting point of sateen and twill will be the same. At the end, fill the free alternate ends and picks with plain weave. Insert

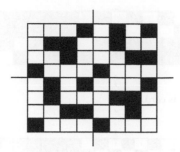

FIGURE 4.59
Crepe (super imposed method).

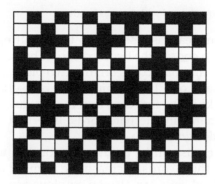

FIGURE 4.60
Crepe (plain based), 7 end sateen.

plain weave in free ends in such a way that every adjacent point to starting point of sateen will have down position. The example in Figure 4.60 shows that seven-end sateen is taken. It requires 14 ends and picks according to alternate ends rule. Then, 3/1+1/3+3/3 twills are inserted into it. At the end plain, weave was filled in free alternate ends.

4.8 Bedford Cord Weave

This is a special class of weave that forms longitudinal warp lines in fabric with fine sunken lines in between. This fabric is used in suiting for ornamental purposes. The method to construct this weave is simple. The repeat of the weave is calculated by multiplying the cord ends by two. The resultant value will be the total number of ends of the weave repeat. The pick repeat is four for this weave. The weave repeat (warp ends) is divided into two halves to construct it. The first and last ends of both the halves are treated

FIGURE 4.61
Bedford cord weave, 10 threads cord.

FIGURE 4.62
Bedford cord weave, 10 threads cord with 2 waded ends.

as cutting ends. Plain weave is inserted on these cutting ends. These plain ends behave as sunken ends in the Bedford cord. Then, simply marks (warp-up) are inserted on in alternate cords on first two picks and on the third and fourth picks of alternate cord. Then, plain weave is inserted above and below (third and fourth picks in the first cord and the first and second picks in the second cord) these marks in alternate cords. Insertion of plain weave will complete the weave design (Figure 4.61).

The width of the cords can be adjusted according to requirement. For more prominence of these cords, increase the number of ends of the cords and vice versa. Wadded ends can also be employed to enhance the prominence of the cords. An equal number of wadded ends are inserted on both sides of the Bedford cord. The position of wadded ends in design must be the same relative to each other on both sides of the design. Wadded ends are those ends that behave same and inserted on regular interval to enhance weave prominence. The design with wadded ends is shown in Figure 4.62.

4.9 Welts and Pique

A pique weave consists of plain face fabric which is composed of a series of warp and weft threads along with a series of stitching threads. This weave is unique due to the formation of horizontal lines (weft wise). This weave requires two beams, one for the plain weave threads and the other for stitching ends. The beam with stitching ends is heavily weighted, but the beam with plain base threads is kept under moderate tension. The stitching ends are woven with the plain threads, and this weaving of stitching end causes the downward pulling of the plain threads. This downward pull causes indentation on the surface of the fabric. To improve the prominence

FIGURE 4.63
Welts and pique weave.

of unstitched portions of the cloth, wadded picks are inserted. The repeat of the weave consists of three ends, while the number of picks depends on the prominence. To get more prominent horizontal lines, the number of picks is increased or specially wadded picks are added.

The word "welt" is concerned to the pique construction, when the indentations make deep or hollow (sunken) lines appear in the cloth. The number of picks per cord varies according to the requirement, but usually the number of consecutive unstitched picks should not exceed from 12. A design of such pique weave is shown in Figure 4.63.

4.10 Mock Leno Weave

This weave is much similar to a gauze-type fabric. The weave is constructed in four quadrants. The first and third quadrants have symmetric weave, and the second and fourth quadrants have opposite weave to the symmetric weave. The perforated fabrics are made by this type of weave. This effect is achieved by reversing the symmetric unit of the weave in the alternate quadrants. So, these weaves are produced in sections that oppose each other. The fabric appearance can be improved or obscured by the system of denting that is employed in this weave. The tendency of threads to run together is counteracted if the last end of one group is passed through the same split as the first end of the next group. The design of mock leno weave is shown in Figure 4.64.

The denting of the above design should be four to avoid the counteracted ends. If denting of three ends is done, then the fourth and fifth ends will counteract each other.

FIGURE 4.64
Mock leno weave.

4.11 Leno Weaving

In leno weaving, the warp yarns are twisted around one another, locking the filling yarns in place. The leno weave is used where relatively a low number of yarns are involved. This is a construction of woven fabrics in which the resulting fabric is very sheer, yet durable. In this weave, two or more warp yarns are twisted around each other as they are interlaced with the filling yarns, thus securing a firm hold on the filling yarn and preventing them from slipping out of position. The yarns work in pairs; one is the standard warp yarn, and the other is the skeleton or doup yarn. This is also called as the gauze weave. Leno weave fabrics are frequently used for window treatments because their structure gives good durability with almost no yarn slippage and permits the passage of light and air. Leno weave improves the stability in "open" fabrics which have a low thread count. Fabrics in leno weave are normally used in conjunction with other weave styles because if used alone, their openness could not produce an effective composite component. The leno weave structure is shown in Figure 4.65.

4.11.1 Leno Woven Structures

Leno weaving is particularly applied where open-effect fabrics are in demand which must have a stable structure. In order to achieve the weave with a nonslip effect, two neighboring warp threads cross over each other. Leno healds, as shown in Figure 4.66, are selected depending on the nature of warp yarn to be woven or the desired opening of shed, and we can select suitable leno healds with lifting healds of synthetic material or stainless steel.

- The leno thread passes through the openings in the guide bar and the stationary ends is drawn into an eyelet in the reed.

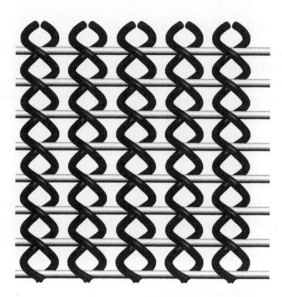

FIGURE 4.65
Leno weave structure.

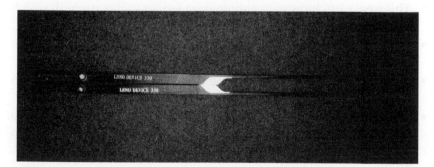

FIGURE 4.66
Leno heald.

- Following weft insertion, the guide bar moves upward and the eyeleted reed downward until the leno thread is over the eyelet of the stationary thread.
- Next, the guide bar is moved sideways until the leno thread is on the other side of the dent in the eyeleted reed.
- Then, the guide bar moves downward and the reed upward, and the shed opens for the next weft insertion.
- As a result of these movements, the leno thread comes to lie over the stationary thread. After insertion of the weft, the motion sequence is repeated in the opposite direction.

4.11.2 Characteristics of Leno Woven Fabrics

- Leno weave has two warp thread systems: ground (basic) and skeleton ends (gauze). By means of two shaft systems—basic and skeleton ends—the skeleton-end threads are led alternately along the left- and right-hand sides of the group of basic threads.

- Ground threads are threaded in the basic harness only; skeleton-end threads are fed into the shaft of the basic harness and to the half-healds of the skeleton-end harness, with the use of special steel wire healds and half-healds.

- The skeleton-end threads are threaded under the ground and fed into the half-healds on the side opposite to that of the basic harness.

- Depending on the wind, we distinguish between half-wound, whole-wound, one-and-a-half-times-wound, twice-wound, and two-and-a-half-times-wound gauzes.

- The gauze weave construction produces a fabric very light in weight and with an open mesh effect.

- Leno weaves also produce curtain materials, some shirting and dress goods.

- This weave produces such light-weight fabrics with a strength which could not be provided by plain weave.

- The gauze weave is sometimes referred to as the leno weave because it is made of a leno loom.

- On the leno loom, the action of one warp yarn is similar to the action of the warp in the plain weave.

- The doup attachment, a hairpin-like device at the heddle, alternately pulls the second warp yarn up or down to the right or left with each pick passage. This causes the pair of warps to be twisted, in effect, around each weft yarn.

- The leno is sometimes used in combination with the plain weave to produce a stripe or figure on a plain background.

- The fabric weight varies depending on the thickness of the yarns, which could be of spun, filament, or combinations of these yarns.

4.11.3 Problems of Leno Woven Fabrics

- A characteristic feature of leno fabrics is that the warp ends do not run parallel but are twisted in pairs or groups between the individual picks.

- The crossing of weft and warp yarns is thus fixed, and the fabric becomes slip resistant. This type of interlacing is achieved using special leno harnesses with appropriate heald.

- Manipulating the harness with the leno heald, drawing in, and repairing of broken ends are very time consuming.
- The yarn and especially also the leno heald are under such high strain that the performance of modern weaving machines cannot be fully exploited.
- Fabric production is correspondingly cost intensive.

4.11.4 Uses of Leno Woven Fabrics

Leno fabrics are extremely tear resistant, of high-quality, easy to process, and applicable in many different areas:

- In the paper and packaging industry, for example, as paper reinforcement for steel packaging products.
- In the building and construction industry as backing material for roofing underlays and for isolation and drainage products.
- As backing fabric in the production of nonwoven and foam products, for example, for molded parts in the automotive industry.
- As spring-interior cover and reinforcement of bolster pieces in the furniture and upholstery industry.
- For laminating applications as backing fabric for different products.

5

Specialty Fabric Structures

Muhammad Umair

CONTENTS

Basic weave designs such as plain, twill, satin, huckaback, crepe, and Bedford cord are most widely used in dresses and other esthetic purposes. But specialty woven fabrics are used for functional as well as esthetic purposes. Different types of specialty woven fabrics such as multilayer woven fabrics, shaped woven, and pile fabrics are available.

5.1 Multilayer Fabric Design

Two-dimensional (2D) fabrics (conventional weave designs) are defined as fabrics having two dimensions. The 2D fabrics are achieved by the interlacement of two sets of yarns (which are perpendicular to each other) in a regular pattern or weaving approach [1]. The strength of fabrics is sustained by the mechanical interlocking of yarns. Three-dimensional (3D) or multilayer woven interlock fabrics are defined as fabrics having substantial thickness, which is achieved by interlacing multilayer warp yarns with the single weft yarn on conventional 2D weaving machines. Multilayer interlock is the structure in which the numbers of threads from different fabric layers are used to bind the layers with each other. A multilayer fabric consists of two or more layers, linked to one another at fastening points whose space from one another in the warp and weft directions is significantly larger than the fundamental weave design.

The 3D fabrics are produced by the interlacement of minimum of three sets of yarn having X (longitudinal), Y (cross), and Z (vertical) directions. The 3D woven fabric is a single fabric structure in which the constituent yarns are theoretically inclined in a three mutually perpendicular directions. The 3D weaving refers to the interlacement of three orthogonal set of yarns on special 3D weaving machines to produce the 3D fabrics [2]. One group of researchers call the multilayer interlock structures as 2.5D structures [2–6] because they do not have a separate Z-direction yarn in the thickness direction while other group of researchers term them as 3D structures because multilayer structures have third dimension in the thickness direction although they use the warp yarns for interlocking [7–11]. So we can call them as 2.5D or multilayer or 3D structures. In 3D fabrics, Z yarn is responsible for strength, stiffness, and thickness of the structure. The 3D woven fabrics have good mechanical properties as compared to 2D fabrics due to their structural stability in through thickness direction and structural integrity. The numbering of layers in 3D multilayer systems is done consecutively from top to bottom. Multilayer woven structures can be produced on conventional weaving machines with little modifications. Multilayer interlock structures are of following types [12]:

- Orthogonal interlock structures
- Angle interlock structures

5.1.1 Orthogonal Interlock Structures

Orthogonal multilayer interlock structures are produced in a way that warp yarns of a layer are used to bind the other layers. Since warp yarns are used to connect the layers, no dedicated binding yarns are used. In orthogonal

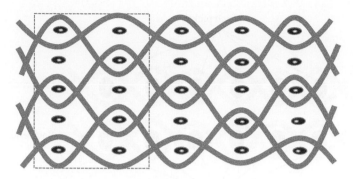

FIGURE 5.1
Multilayer orthogonal layer-to-layer structures.

layer-to-layer structures, certain warp yarns are used to bind two connective layers together and so on, as shown in Figure 5.1. In orthogonal through thickness structures, some warp yarns from the first and last layers are used to bind all the layers present in the fabric, as shown in Figure 5.2.

5.1.1.1 Weave Designs of Orthogonal Layer-to-Layer Stitched Structures

Fabrics consisting of two or more single-layer, connected to each other at some points whose distance from each other in the warp and weft directions is substantially greater than the basic weave repeat are called multilayer stitched fabrics. An example of two-layer interlock fabrics is shown in Figure 5.3. This shows the weaving of a two-layer fabric; the individual weave of each fabric layer is plain. It is the simplest example of a multilayer fabric.

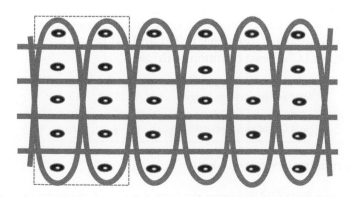

FIGURE 5.2
Multilayer orthogonal through thickness.

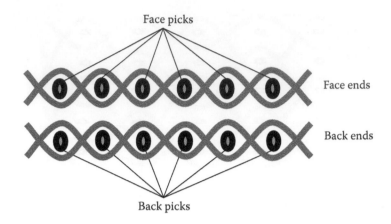

FIGURE 5.3
Cross-sectional view of two-layer structure.

The first (top) layer is called the face (F) and the second (bottom) is called the Back (B) layer. On a graph paper, these are marked with different signs, as shown in the following:

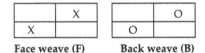

Face weave (F)   **Back weave (B)**

In the resultant design of the multilayer fabrics, the ends are arranged in face-back-face-back order and picks are also arranged in the same face-back-face-back sequence, resulting in a two-layer tubular fabric. The repeat of this design is completed on 4 ends × 4 picks.

Repeat Size

Repeat size = R = LCM of weave designs × number of weave designs.
We have two weave designs, that is, 1/1 and 1/1.
Therefore, for LCM (least common multiple)

We have two weave designs	=	1 + 1	1 + 1
	=	2	2
So, LCM of two weave designs	=	**2**	

And the number of weave designs = 2.
Finally,
Repeat size = R = LCM of weave designs × number of weave designs.
Repeat size = R = 2 × 2.
Repeat size = R = 4.

The complete weave design will be on 4 ends and 4 picks as given below.

	F	B	F	B
B				
F				
B				
F				

Rules of interlacement for a multilayer weave are described as follows:
Rule 1: Face ends will only interlace with face picks.

	F	B	F	B
B				
F	■		X	
B				
F	X		■	

X = face ends raised on face picks
■ = face ends down on face picks

Rule 2: Back ends will only weave with back picks.

	F	B	F	B
B		■		O
F				
B		O		■
F				

O = back ends raised on back picks
■ = back ends down on back picks

Combining face and back fabric designs, we have

	F	B	F	B
B		■		O
F	■		X	
B		O		■
F	X		■	

It seems that the design is complete as the weave of face with face and back with back is complete, which is required in the resultant fabric. But still the loom *"does not know"* how to split the two layers apart. For the weaving machine, the design is as follows:

			X
		X	
	X		
X			

1/3 Z twill

To weave it in two separate layers, we will have to tell the machine to separate it into two layers.

Rule 3: All the face ends should be raised on all the back picks.

B	A	■	A	O
F	■		X	
B	A	O	A	■
F	X		■	
	F	B	F	B

A = face ends raised on back picks

So, the resultant two-layer fabric design without any stitching for the weaving machine would be

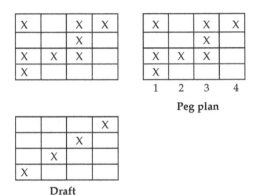

Peg plan

Draft

This stitching forms the foundation of 3D weaving. We can stitch a number of layers together forming a very thick 3D fabric. There are three different techniques used for stitching different layers together: raiser stitching, sinker stitching, and extra end stitching.

Raiser stitching means certain back layer ends are raised over certain face picks as shown in Figure 5.4.

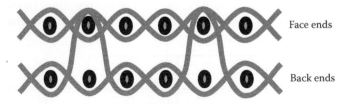

FIGURE 5.4
Cross-sectional view of raiser stitching of two layers.

The resultant weave design of orthogonal layer-to-layer structures having raiser stitching would be

	F	B	F	B
B	A	■	A	O
F	■		X	
B	A	O	A	■
F	X		■	S

Without any stitching

	F	B	F	B
B	A	■	A	O
F	■	S	X	S
B	A	O	A	■
F	X	S	■	S

With full stitching points
S = back ends raised from the front picks

	F	B	F	B
B	A	■	A	O
F	■		X	S
B	A	O	A	■
F	X	S	■	

Alternate stitching points
S = Back ends raised from the front picks

Sinker stitching means certain face layer ends are lowered under certain back picks as shown in Figure 5.5.

So, the resultant weave design of orthogonal layer-to-layer structures having sinker stitching would be

	F	B	F	B
B	A	■	A	O
F	■		X	
B	A	O	A	■
F	X		■	

Without any stitching

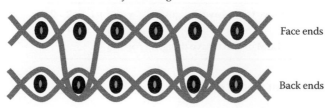

Face ends

Back ends

FIGURE 5.5
Cross-sectional view of sinker stitching of two layers.

B	A	■	C	O
F	■		X	
B	C	O	A	■
F	X		■	
	F	B	F	B

A	■		O
■		X	
	O	A	■
X		■	

With alternate stitching points (left) and equivalent design (right)
C = face ends passed under the back picks (on this pick, end will be
    down in weave design)

Extra end stitching uses additional warp threads to stitch all the layers together. There is no rule as to the number of ends that will be used for binding, nor how often these ends should stitch the two layers together [13].

The weave design of three-layered orthogonal layer-to-layer raiser stitched structures having plain weave in every layer is as follows (repeat size = 6 × 6):

B	D	E	■	D	E	C
M	D	■	S	D	F	S
F	■	S		A	S	
B	D	E	C	D	E	■
M	D	F	S	D	■	S
F	A	S		■	S	
	F	M	B	F	M	B

Note: All alphabets show that warp ends are passing over the running picks; S alphabet shows the stitching points.

The weave design of two-layered orthogonal layer-to-layer raiser stitched structures having 3/1 twill and 2/2 twill weaves in front and back layers, respectively, is given in the following:

Repeat Size

Repeat size = R = LCM of weave designs × number of weave designs.
We have two weave designs, that is, 3/1 and 2/1.
Therefore, for LCM

We have two weave designs	=	3 + 1	2 + 2
	=	4	4
So, LCM of two weave designs	=	4	

And the number of weave designs = 2.
Finally,
Repeat size = R = LCM of weave designs × number of weave designs.
Repeat size = R = 4 × 2.

Repeat size $= R = 8$ and complete weave design will be on 8 ends and 8 picks as given in the following:

B	E	■	E	■	E	D	E	D
F	■		A	S	A		A	S
B	E	■	E	D	E	D	E	■
F	A	S	A		A	S	■	
B	E	D	E	D	E	■	E	■
F	A		A	S	■		A	S
B	E	D	E	■	E	■	E	D
F	A	S	■		A	S	A	
	F	B	F	B	F	B	F	B

Note: All alphabets show that warp ends are passing over the picks; S alphabet shows the stitching points.

5.1.1.2 Weave Designs of Orthogonal Through Thickness Stitched Structures

In orthogonal through thickness structures, some warp yarns from the first and last layers are used to bind all the layers present in the fabric. Formula to calculate the repeat size of orthogonal through thickness stitched structures is same as that of orthogonal layer-to-layer stitched structures, but having two extra ends for through thickness stitching.

The weave design of two-layered orthogonal through thickness stitched structure having 3/1 twill and 2/2 twill weaves in front and back layers, respectively, is given in the following (repeat size will be on 10 ends and 8 picks):

B	E	■	E	■	E	D	E	D		X
F	■		A		A		A			X
B	E	■	E	D	E	D	E	■		X
F	A		A		A		■			X
B	E	D	E	D	E	■	E	■	X	
F	A		A		■		A		X	
B	E	D	E	■	E	■	E	D	X	
F	A		■		A		A		X	
	F	B	F	B	F	B	F	B	S1	S2

Note: All alphabets show that warp ends are passing over the running picks.

Two extra ends (S1 and S2) are used for through thickness stitching of all the layers together. The weave design of three-layered orthogonal through thickness stitched structures having plain weave in every layer is as follows (repeat size $= 8$ ends $\times 6$ picks):

	F	M	B	F	M	B	S1	S2
B	D	E	■	D	E	C		X
M	D	■		D	F			X
F	■			A				X
B	D	E	C	D	E	■	X	
M	D	F		D	■		X	
F	A			■			X	

Note: All alphabets show that warp ends are passing over the running picks.

The weave design of six-layered orthogonal through thickness stitched structures having plain weave in every layer is as follows (repeat size = 14 ends × 12 picks):

	F1	F2	F3	B3	B2	B1	F1	F2	F3	B3	B2	B1	S1	S2
B1	X	X	X	X	X		X	X	X	X	X	X		X
B2	X	X	X	X			X	X	X	X	X			X
B3	X	X	X				X	X	X	X				X
F3	X	X					X	X	X					X
F2	X						X	X						X
F1							X							X
B1	X	X	X	X	X	X	X	X	X	X	X		X	
B2	X	X	X	X	X		X	X	X	X			X	
B3	X	X	X	X			X	X	X				X	
F3	X	X	X				X	X					X	
F2	X	X					X						X	
F1	X												X	

5.1.2 Angle Interlock Structures

Multilayer angle interlock structures are produced in a way that binding yarns passed from the fabric layers at a certain angle to bind all the layers. Angle interlock structures mainly are of two types: through thickness and layer-to-layer angle interlock structures. The layer-to-layer angle interlock is a multilayer preform in which warp yarn travels from one layer to the next layer, and back as shown in Figure 5.6. But the through thickness angle interlock is a multilayered preform in which warp yarn travels from one face of the structure to the other, binding mutually all the layers of the preforms as shown in Figure 5.7.

5.1.2.1 Weave Designs of Angle Interlock Layer-to-Layer Stitched Structures

In orthogonal structures, the weave design could be drawn just by knowing the number of layers in the structure. But in angle interlock structures, first we have to draw the cross section of the design and the repeat size should

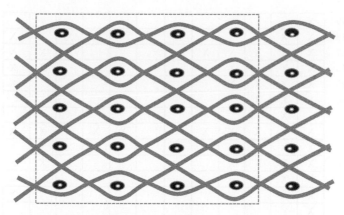

FIGURE 5.6
Multilayer angle interlock layer-to-layer structure.

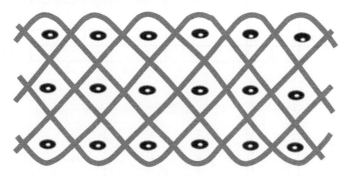

FIGURE 5.7
Multilayer angle interlock through thickness structure.

be marked from the cross section. Then in cross section, the numbering of layers in warp and weft directions is done.

For example, we want to make a six-layered layer-to-layer stitched angle interlock structure. First, we draw the cross section of the fabric design and the numbering of layers is done both in warp and weft directions, as shown in Figure 5.8.

Keeping in view the cross section of the design, the weave design of six-layered layer-to-layer stitched angle interlock structures could be drawn on a graph paper, which is given as follows:

24	X	X	X	X	X	X		X	X	X	X	X
23	X	X	X	X	X				X	X	X	X
22	X	X	X	X						X	X	X
21	X	X	X								X	X
20	X	X										X

	1	2	3	4	5	6	7	8	9	10	11	12
19	X											
18	X	X	X	X	X	X	X		X	X	X	X
17	X	X	X	X	X	X				X	X	X
16	X	X	X	X	X						X	X
15	X	X	X	X								X
14	X	X	X									
13		X										
12	X	X	X	X	X	X		X	X	X	X	X
11	X	X	X	X	X				X	X	X	X
10	X	X	X	X					X	X	X	X
9	X	X	X								X	X
8	X	X										X
7	X											
6	X	X	X	X	X		X	X	X	X	X	X
5	X	X	X	X				X	X	X	X	X
4	X	X	X						X	X	X	X
3	X	X								X	X	X
2	X										X	X
1												X

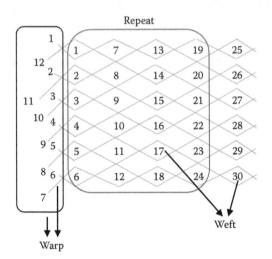

FIGURE 5.8
Cross section of six-layered layer-to-layer stitched angle interlock structure.

5.1.2.2 Weave Designs of Angle Interlock Through Thickness Stitched Structures

Like layer-to-layer angle interlock structures, first we draw the cross section of the design and secondly the numbering of layers is made both in warp and weft directions for through thickness angle interlock structures.

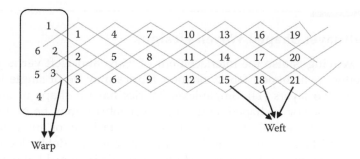

FIGURE 5.9
Cross section of three-layered through thickness stitched angle interlock structure.

For example, we want to make a three-layered through thickness stitched angle interlock structure. First, we draw the cross section of the fabric design and the numbering of layers is done both in warp and weft directions, as shown in Figure 5.9.

Keeping in view the cross section of the design, the weave design of a three-layered through thickness stitched angle interlock structure could be drawn on a graph paper, which is given as follows:

	1	2	3	4	5	6
21	X	X		X	X	X
20	X				X	X
19						X
18	X	X	X		X	X
17	X	X				X
16	X					
15	X	X	X	X		X
14	X	X	X			
13		X				
12	X	X	X	X	X	
11		X	X	X		
10			X			
9		X	X	X	X	X
8			X	X	X	
7				X		
6	X			X	X	X
5				X	X	X
4					X	
3	X	X		X	X	X
2	X				X	X
1						X

5.2 Multilayer Shaped Woven Designs

Shaped weaving is also the interlacement of warp and weft yarns, but the final product is a shaped fabric. Shaped fabrics may be of single layer or more than two layers [14]. Single-layer fabrics have a shape of any type such as a corner or V-shaped, and multilayer fabrics may be of T-shaped or H-shaped, tetrahedral shape, seamless woven fashion, or any hole in the fabric. Shaped fabrics can be produced on conventional dobby looms with no extra atomization in the loom and on electronic jacquard loom and also on 3D loom, but this is not commonly available. As woven fabrics have excellent properties such as stiffness, strength, and dimensional stability, they are used in technical textiles. Pile yarns are added to provide sufficient stability and compression. Spacer fabrics are used in air and space industries, tires of vehicles, and automotive and transportation industries to achieve good tensile strength, compression, and stiffness; the material properties and process properties are optically combined [15].

There are different techniques that are used to produce shaped parts. Some of these are as follows:

- The interlacement of two sets of yarns to produce a 3D cloth on a conventional loom.
- Interlacing three yarns (ground warp, pile warp, and pile weft) produces pile fabrics.
- The actual 3D weaving method interlaces three orthogonal sets of yarns. It produces a totally interlaced 3D cloth, on a specifically designed 3D weaving machine [8].

Jacquard weaving machines help to give different integrated structures in one piece of fabric.

Multilayer shaped woven structures can be categorized as

- Closed structures
- Open structures

5.2.1 Closed Structures

Tubular fabric structures can be produced by weaving, knitting, and braiding. Closed or tubular woven fabrics are the jointless or seamless circular structures, produced on that type of loom which has a continuous pick insertion mechanism like shuttle loom. On shuttleless looms, that is, air-jet, projectile, and rapier, tubular structures can be produced by stitching at the selvedge points. But in the second type of tubular structures, strength at the stitched selvedge point will not be even. Tubular woven structures can be prepared using both layer-to-layer and through thickness stitching techniques.

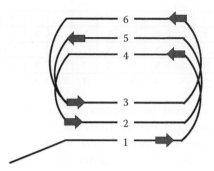

FIGURE 5.10
Cross section of three-layered tubular structure.

For example, three-layered tubular structure is formed with layer-to-layer raiser stitching having every layer plain weave. So it has three face layers and three back layers. Face layers are designated by 6, 5, and 4, and back layers by 3, 2, and 1, as shown in Figure 5.10. For raiser stitching, Layer 4 is stitched with Layer 5 and Layer 5 with Layer 6. Similarly, Layer 1 is stitched with Layer 2 and Layer 2 is stitched with Layer 3. In order to make a tubular structure, Layer 3 is not stitched with Layer 4.

Repeat Size

Repeat size = R = LCM of weave designs × number of weave designs.
Therefore, for LCM

We have six weave designs	=	$1+1$	$1+1$	$1+1$	$1+1$	$1+1$	$1+1$
	=	2	2	2	2	2	2
So, LCM of six weave designs	=	2					

And the number of weave designs = 3.
Finally,
Repeat size = R = LCM of weave designs × number of weave designs.
Repeat size = R = 2 × 6.
Repeat size = R = 12.

The complete weave design will be on 12 ends and 12 picks. So, the weave design of a three-layered tubular structure is given below:

6					S						S	X
3		S		X	X	X		S	X	X	X	X
6					S	X					S	
3		S	X	X	X	X		S		X	X	X
5				S		X				S	X	X
2	S			X	X	X	S	X	X	X	X	X
5				S	X	X				S		X
2	S	X	X	X	X	X	S		X	X	X	X

	1	2	3	4	5	6	1	2	3	4	5	6
4					X	X				X	X	X
1		X	X	X	X	X	X	X	X	X	X	X
4			X	X	X						X	X
1	X	X	X	X	X	X		X	X	X	X	X

Furthermore, if we want to prepare a three-layered hybrid material tubular structure having different material in each layer, then we take shuttle loom with drop box (multi-weft insertion) assembly for different materials picking, as shown in Figure 5.11, but the weave design will remain same.

So, the weave design with different materials in warp and weft directions is given below. C1 and C4 represent the one type of material (e.g., cotton), P2 and P5 represent the second type of material (i.e., polyester), and V3 and V6 represent the third type of material (i.e., viscose) both in warp and weft directions. So, we need three different shuttles to insert three different types of materials in the weft.

	C1	P2	V3	C4	P5	V6	C1	P2	V3	C4	P5	V6
V6					S						S	X
V3		S		X	X	X		S	X	X	X	X
V6					S	X					S	
V3		S	X	X	X	X		S		X	X	X
P5				S		X				S	X	X
P2	S		X	X	X	X	S	X	X	X	X	X
P5				S	X	X				S		X
P2	S	X	X	X	X	X	S		X	X	X	X
C4					X	X				X	X	X
C1		X	X	X	X	X	X	X	X	X	X	X
C4				X	X	X					X	X
C1	X	X	X	X	X	X		X	X	X	X	X

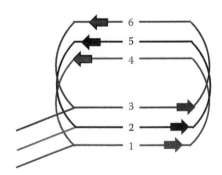

FIGURE 5.11
Cross section of three-layered hybrid tubular structure.

Similarly, a three-layered tubular structure (Figure 5.10) can be produced with the through thickness orthogonal stitching with simple and hybrid materials. Repeat size for the three-layered tubular structure having the through thickness orthogonal stitching is 16 ends (12 + 2 + 2 extra ends for through thickness stitching) and 12 picks. The weave design is given below:

	1	2	3	4	5	6	1	2	3	4	5	6	S1	S2	S3	S4
6													X			X
3			X	X	X			X	X	X	X				X	X	X
6				X											X	
3			X	X	X	X			X	X	X	X			X	X
5				X						X	X					X
2			X	X	X	X	X	X	X	X	X				X	X	X
5				X	X						X				X	
2		X	X	X	X	X		X	X	X	X	X				X	X
4				X	X					X	X	X				X
1		X	X	X	X	X	X	X	X	X	X	X		X	X	X
4				X	X	X				X	X				X	
1	X	X	X	X	X	X		X	X	X	X	X	X		X	X

5.2.1.1 Nodal Structure

Another type of closed structure is nodal structure consisting of a long circular tube, which is further divided into two or three similar tubes, as shown in Figure 5.12. Nodal structures are produced on a continuous weft insertion (shuttle) loom with dobby mechanism having drop box assembly (multiple weft shuttles).

Nodal structure is divided into two portions: the first part is the long main trunk and the second part is the branching portion. We use different weave designs for both portions of node. For the first part of node, that is, main trunk, we use the single shuttle and follow the below weave design of a four-layered tubular structure without any stitching having plain weave in each layer.

	1	2	3	4	1	2	3	4
4	X	X	X		X	X	X	X
3	X	X			X	X	X	
2	X				X	X		
1					X			
4	X	X	X	X	X	X	X	
3	X	X	X		X	X		
2	X	X			X			
1	X							

The above weave design is produced keeping in view the layers arrangement (L1, L2, L3, and L4 are four layers), as shown in Figure 5.13, which after

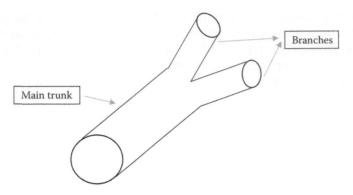

FIGURE 5.12
Nodal structure.

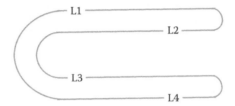

FIGURE 5.13
Cross section of designing for main trunk part.

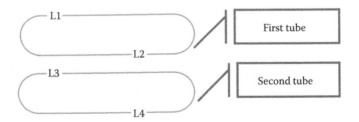

FIGURE 5.14
Cross section of designing for branches.

opening or in an off-loom condition produces the tube-like main trunk of the nodal structure.

For the second branching part of the nodal structure, two different shuttles will be used for the two separate tubes. The first shuttle produces the first top tube and the second shuttle produces the second or bottom tube of the structure, as shown in Figure 5.14.

The design for the two separate tubes using two shuttles is given below. The highlighted portion produces the first tube using the first shuttle and the un-highlighted portion produces the second tube using the other shuttle.

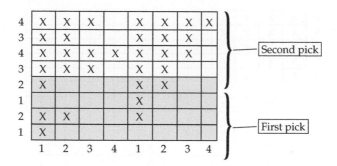

	1	2	3	4	1	2	3	4	
4	X	X	X		X	X	X	X	⎫
3	X	X			X	X	X		Second pick
4	X	X	X	X	X	X	X		
3	X	X	X		X	X			⎬
2	X				X	X			
1					X				
2	X	X			X				First pick
1	X								⎭

5.2.1.2 Wheel Shape

Another example of closed shape that we can weave by multilayer weaving on a continuous picking loom (Shuttle) is wheel. The cross-sectional view of the wheel design is shown in Figure 5.15. Six-layered orthogonal layer-to-layer raiser stitched structures with plain weave in each layer consist of five separate parts due to different stitching portions.

	1	2	3	4	5	6	1	2	3	4	5	6
6	X	X	X	X	X		X	X	X	X	X	X
5	X	X	X	X		S	X	X	X	X	X	S
4	X	X	X				X	X	X	X		
3	X	X		S			X	X	X	S		
2	X						X	X				
1		S					X	S				
6	X	X	X	X	X	X	X	X	X	X	X	
5	X	X	X	X	X	S	X	X	X	X		S
4	X	X	X	X			X	X	X			
3	X	X	X	S			X	X		S		
2	X	X					X					
1	X	S					S					

Part A

	1	2	3	4	5	6	1	2	3	4	5	6
6	X	X	X	X	X		X	X	X	X	X	X
5	X	X	X	X			X	X	X	X	X	
4	X	X	X				X	X	X	X		
3	X	X					X	X	X			
2	X						X	X				
1							X					
6	X	X	X	X	X	X	X	X	X	X	X	
5	X	X	X	X	X		X	X	X	X		
4	X	X	X	X			X	X	X			
3	X	X	X				X	X				
2	X	X					X					
1	X											

Part B

	1	2	3	4	5	6	1	2	3	4	5	6
6	X	X	X	X	X		X	X	X	X	X	X
5	X	X	X	X			X	X	X	X	X	
4	X	X	X		S		X	X	X	X	S	
3	X	X					X	X	X			
2	X		S				X	X	S			
1							X					
6	X	X	X	X	X	X	X	X	X	X	X	
5	X	X	X	X	X		X	X	X	X		
4	X	X	X	X	S		X	X	X		S	
3	X	X	X				X	X				
2	X	X	S				X			S		
1	X											
	1	2	3	4	5	6	1	2	3	4	5	6

Part C

5.2.2 Open Structures

Open multilayer structures having any specific shape can be produced on conventional looms. Multilayer open structures have more than one layer, which may or may not be connected at some points. The fabrics in which layers may be self-stitched are known as interlocked fabrics. Shaped woven fabrics can be produced on conventional dobby looms. Different shapes such as H, I, hemispherical, tetrahedron, and pyramidal can be used in angled parts of machines and in construction industry as well.

5.2.2.1 T and H Shapes

T and H shapes (Figure 5.16) with a different stitching pattern could be produced on conventional dobby looms.

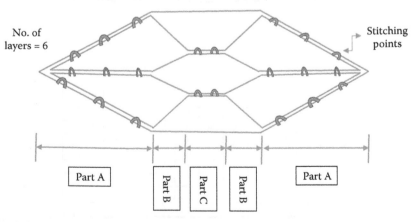

FIGURE 5.15
Cross-sectional view of wheel design.

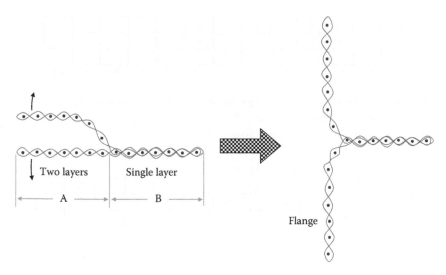

FIGURE 5.16
Cross section of T-shaped fabrics.

Multilayer (3D) techniques are used to produce the shaped woven fabrics on conventional looms. For the production of T-shaped fabric on loom, we select the orthogonal layer-to-layer stitching pattern in this part. T is divided into two portions: the first is the main trunk and the second is the branches of T, as shown in Figure 5.16. The weave design of the main trunk of T and the second part will be different to achieve the required shape. Fabrics produced from loom will give us T shape in an off-loom condition. For the first part of T shape, six layers will be stitched together to produce the main trunk, and for the second part of T shape (branches), top three and bottom three layers will be stitched together to produce two branches of T. The weave design for the six-layered T shaped fabric with the orthogonal layer-to-layer raiser stitching having plain weave in each layer is given below.

Design for branches of T shape (Part A)

F1, F2, and F3 are stitched together
B1, B2, and B3 are stitched together

B1	X	X	X	X	X		X	X	X	X	X	X
B2	X	X	X	X		S	X	X	X	X	X	
B3	X	X	X				X	X	X	X	S	
F3	X	X					X	X	X			
F2	X						X	X	S			
F1		S					X					
B1	X	X	X	X	X	X	X	X	X	X	X	
B2	X	X	X	X	X		X	X	X	X		S
B3	X	X	X	X	S		X	X	X			

F3	X	X	X				X	X				
F2	X	X	S				X					
F1	X							S				
	F1	F2	F3	B3	B2	B1	F1	F2	F3	B3	B2	B1

The cross section of Part A after splitting the six layers into two equal groups of layers is shown in Figure 5.17.

Design for the main trunk of T shape (Part B)

Six layers are stitched together

B1	X	X	X	X	X		X	X	X	X	X	X
B2	X	X	X	X		S	X	X	X	X	X	
B3	X	X	X				X	X	X	X	S	
F3	X	X		S			X	X	X			
F2	X						X	X	S			
F1		S					X					
B1	X	X	X	X	X	X	X	X	X	X	X	
B2	X	X	X	X	X		X	X	X	X		S
B3	X	X	X	X	S		X	X	X			
F3	X	X	X				X	X		S		
F2	X	X	S				X					
F1	X							S				
	F1	F2	F3	B3	B2	B1	F1	F2	F3	B3	B2	B1

So, the final peg plan to produce orthogonal layer-to-layer stitched shaped fabrics is given below:

12	X	X	X	X	X		X	X	X	X	X	X	X	X
11	X	X	X	X		S	X	X	X	X	X		X	X
10	X	X	X				X	X	X	X	S			X
9	X	X		S			X	X	X					
8	X						X	X	S					
7		S					X							
6	X	X	X	X	X	X	X	X	X	X	X		X	X
5	X	X	X	X	X		X	X	X	X		S	X	X
4	X	X	X	X	S		X	X	X				X	
3	X	X	X				X	X		S				
2	X	X	S				X							
1	X							S						
	1	2	3	4	5	6	7	8	9	10	11	12	13	14

For T-shaped fabrics during drawing in pattern, total ends in the weavers beam will be divided into two parts as given below:

Ends of Part A + Ends of Part B = T shape.

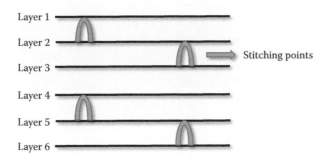

FIGURE 5.17
Stitching cross section of part A.

We can achieve more number of T shapes from single weavers beam by dividing the beam into more parts:

{Ends of Part A + Ends of Part B} + {Ends of Part A + Ends of Part B} = Two T shapes.

For H-shaped fabrics during drawing in pattern, total ends in the weavers beam will be divided into three parts as given below:

Ends of Part A + Ends of Part B + Ends of Part A = H shape.

We can achieve more number of H shapes from single weavers beam by dividing the beam into more parts:

{Ends of Part A + Ends of Part B + Ends of Part A} + {Ends of Part A + Ends of Part B + Ends of Part A} = Two H shapes.

Similarly, we can produce the T-shaped fabric with the orthogonal through thickness stitching. Same rule will be followed like that of orthogonal layer-to-layer raiser stitching, but the stitching technique will be different. The weave design for the six-layered T-shaped fabric with the orthogonal through thickness stitching having plain weave design in each layer is given below:

Design for branches of T shape (Part A)

F1, F2, and F3 are stitched together
B1, B2, and B3 are stitched together

	F1	F2	F3	B3	B2	B1	F1	F2	F3	B3	B2	B1	S1	S2	S3	S4
B1	X	X	X	X	X		X	X	X	X	X	X	X	X	X	
B2	X	X	X	X			X	X	X	X	X		X	X	X	
B3	X	X	X				X	X	X	X			X	X	X	
F3	X	X					X	X	X				X			
F2	X						X	X					X			
F1							X						X			
B1	X	X	X	X	X	X	X	X	X	X	X		X	X		X
B2	X	X	X	X	X		X	X	X	X			X	X		X
B3	X	X	X	X			X	X	X				X	X		X
F3	X	X	X				X	X						X		
F2	X	X					X							X		
F1	X						X							X		

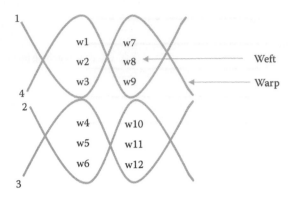

FIGURE 5.18
Stitching cross section of part A.

The cross section of Part A after splitting the six layers into two equal groups of layers is shown in Figure 5.18.

Design for the the main trunk of T shape (Part B)

Six layers are stitched together

	F1	F2	F3	B3	B2	B1	F1	F2	F3	B3	B2	B1	S5	S6
B1	X	X	X	X	X		X	X	X	X	X	X		X
B2	X	X	X	X			X	X	X	X	X			X
B3	X	X	X				X	X	X	X				X
F3	X	X					X	X	X					X
F2	X						X	X						X
F1							X							X
B1	X	X	X	X	X	X	X	X	X	X	X		X	
B2	X	X	X	X	X		X	X	X	X			X	
B3	X	X	X	X			X	X	X				X	
F3	X	X	X				X	X					X	
F2	X	X					X						X	
F1	X												X	

So, the final peg plan to produce the orthogonal layer-to-layer stitched shaped fabric is given below:

12	X	X	X	X	X		X	X	X	X	X	X	X	X	X		X
11	X	X	X	X			X	X	X	X	X		X	X	X		X
10	X	X	X				X	X	X	X			X	X	X		X
9	X	X					X	X	X				X				X
8	X						X	X					X				X

7							X						X					X
6	X	X	X	X	X	X	X	X	X	X	X		X	X		X	X	
5	X	X	X	X	X		X	X	X	X			X	X		X	X	
4	X	X	X	X			X	X	X				X	X		X	X	
3	X	X	X				X	X					X			X		
2	X	X					X						X			X		
1	X												X			X		
	1	2	3	4	5	6	7	8	9	10	11	12	13	14	15	15	17	18

For T-shaped fabrics during drawing in pattern, total ends in the weavers beam will be divided into two parts as given below:

Ends of Part A + Ends of Part B = T shape.

We can achieve more number of T shapes from single weavers beam by dividing the beam into more parts:

For H-shaped fabrics during drawing in pattern, total ends in the weavers beam will be divided into three parts as given below:

Ends of Part A + Ends of Part B + Ends of Part A = H shape.

We can achieve more number of H shapes from single weavers beam by dividing the beam into more parts:

Umair et al. [16] compared the mechanical properties of orthogonal layer-to-layer and through thickness stitched T- and H-shaped woven three-dimensional fabrics and concluded that the mechanical behavior (peel off strength and breaking load) of layer-to-layer interlocked structures in T and H shapes was found to be better than that of through thickness structures.

5.2.2.2 Turbine Blade Shape

Another shape that we can weave by multilayer weaving is turbine blade. The cross-sectional view of the turbine blade design is shown in Figure 5.19. Four-layered orthogonal layer-to-layer raiser stitched structures with plain weave in each layer consist of five separate parts due to different stitching portions.

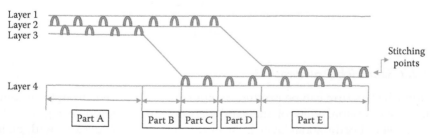

FIGURE 5.19
Cross-sectional view of turbine blade design.

The weave design for five different parts of turbine blade is given below:

4	X	X	X		X	X	X	X	X	X	X		X	X	X	X
3	X	X			X	X	X		X	X			X	X	X	
2	X		S		X	X	S		X				X	X		
1		S			X	S				S			X	S		
4	X	X	X	X	X	X	X		X	X	X	X	X	X	X	
3	X	X	X		X	X			X	X	X		X	X		
2	X	X	S		X		S		X	X			X			
1	X	S				S			X	S				S		
	1	2	3	4	1	2	3	4	1	2	3	4	1	2	3	4

Part A (columns 1–8) Part B (columns 9–16)

4	X	X	X		X	X	X	X	X	X	X		X	X	X	X
3	X	X		S	X	X	X	S	X	X		S	X	X	X	S
2	X				X	X			X				X	X		
1		S			X	S							X			
4	X	X	X	X	X	X	X		X	X	X	X	X	X	X	
3	X	X	X	S	X	X		S	X	X	X	S	X	X		S
2	X	X			X				X	X			X			
1	X	S				S			X							
	1	2	3	4	1	2	3	4	1	2	3	4	1	2	3	4

Part C (columns 1–8) Part D (columns 9–16)

4	X	X	X		X	X	X	X
3	X	X		S	X	X	X	S
2	X		S		X	X	S	
1					X			
4	X	X	X	X	X	X	X	
3	X	X	X	S	X	X		S
2	X	X	S		X		S	
1	X							
	1	2	3	4	1	2	3	4

Part E

5.2.2.3 Sandwich Structure

Sandwich structure can be produced using a multilayer weaving technique. The cross-sectional view of the sandwich design is shown in Figure 5.20. Seven-layered orthogonal layer-to-layer raiser stitched structures with plain weave in each layer consist of three separate parts due to different stitching portions.

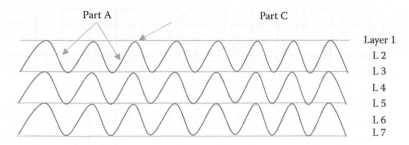

FIGURE 5.20
Cross-sectional view of sandwich structure.

The weave design for three different parts of sandwich structure is given below. The complete weave design will consist of the combination of three parts.

	1	2	3	4	5	6	7	8	9	10	11	12	13	14
7	X	X	X	X	X	X		X	X	X	X	X	X	X
6	X	X	X	X	X			X	X	X	X	X	X	
5	X	X	X	X				X	X	X	X	X		
4	X	X	X					X	X	X	X			
3	X	X						X	X	X				
2	X							X	X					
1								X						
7	X	X	X	X	X	X	X	X	X	X	X	X	X	
6	X	X	X	X	X	X		X	X	X	X	X		
5	X	X	X	X	X			X	X	X	X			
4	X	X	X	X				X	X	X				
3	X	X	X					X	X					
2	X	X						X						
1	X													

Part A

	1	2	3	4	5	6	7	8	9	10	11	12	13	14
7	X	X	X	X	X	X		X	X	X	X	X	X	X
6	X	X	X	X	X			X	X	X	X	X	X	
5	X	X	X	X		S		X	X	X	X	X		
4	X	X	X					X	X	X	X			
3	X	X		S				X	X	X				
2	X							X	X					
1		S						X						
7	X	X	X	X	X	X	X	X	X	X	X	X	X	
6	X	X	X	X	X	X		X	X	X	X	X		
5	X	X	X	X	X			X	X	X	X			
4	X	X	X	X				X	X	X				

Part B

3	X	X	X					X	X					
2	X	X						X						
1	X													
	1	2	3	4	5	6	7	8	9	10	11	12	13	14

Part C

7	X	X	X	X	X	X		X	X	X	X	X	X	X
6	X	X	X	X	X		S	X	X	X	X	X	X	
5	X	X	X	X				X	X	X	X	X		
4	X	X	X		S			X	X	X	X			
3	X	X						X	X	X				
2	X		S					X	X					
1								X						
7	X	X	X	X	X	X	X	X	X	X	X	X	X	
6	X	X	X	X	X	X		X	X	X	X	X		
5	X	X	X	X	X			X	X	X	X			
4	X	X	X	X				X	X	X				
3	X	X	X					X	X					
2	X	X						X						
1	X													
	1	2	3	4	5	6	7	8	9	10	11	12	13	14

5.2.2.4 Spacer Fabric

Spacer fabrics are fabrics in which two outer layers of fabrics are joined together by means of spacer yarn through crosslinking and also through pile yarn [17], as shown in Figure 5.21. Spacer fabrics have compression property,

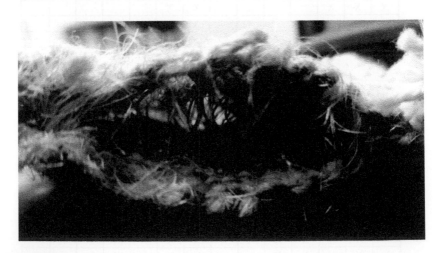

FIGURE 5.21
Spacer fabric.

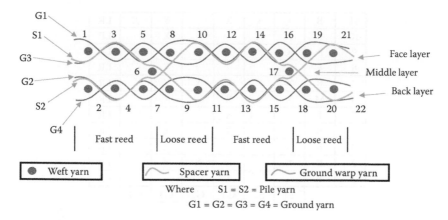

FIGURE 5.22
Cross-sectional view of a spacer fabric design.

impact resistance, thermal insulation, thermal conductivity, air permeability, and absorbency.

Due to their better air permeability, spacer fabrics are widely used in sports clothing, and because of their good absorbency, they are used in medical textiles. Spacer fabrics are also used in technical textiles and in geo textiles as a reinforcement material between aggregate and soil stone and in roads, railways works, erosion prevention, and separation.

The multilayer weave of spacer fabrics consists of three sets of yarn including warp yarn, weft yarn, and spacer/pile yarn. The cross-sectional view of a multilayer spacer fabric is shown in Figure 5.22. The purpose of loose reed is to collect a certain number of picks unbeaten and then beaten through fast reed mechanism to get compact picks on a terry loom.

The weave design of a multilayer space fabric, as per the cross section shown in Figure 5.22, is given below:

22	B	X		X		X	X	LR
21	F	X				X		LR
20	B	X	X	X		X	X	LR
19	F					X		LR
18	B	X	X	X	X	X		FR
17	M	X		X		X		FR
16	F			X				FR
15	B		X	X		X	X	FR
14	F		X			X		FR
13	B	X	X	X	X	X		FR
12	F			X				FR

11	B		X	X		X	X	LR
10	F		X			X		LR
9	B	X	X	X		X	X	LR
8	F					X		LR
7	B	X	X	X	X	X		FR
6	M	X		X		X		FR
5	F			X				FR
4	B	X		X		X	X	FR
3	F	X				X		FR
2	B	X	X	X	X	X		FR
1	F			X				FR
		S1	S2	G1	G2	G3	G4	

where
LR = lose reed,
FR = fast reed,
B = back layer,
M = middle layer,
F = face layer,
S1 = S2 = pile yarn,
G1 = G2 = G3 = G4 = ground yarn.

5.3 Non-Crimped Fabric (3D Non-Crimp Fabric)

A non-crimp fabric (NCF) is a fabric that has all warp and weft yarns practically straight [18]. The NCF ideally has zero crimp in it, but in actual case, it has a negligible crimp (both or any one direction). The basic principle that leads toward innovation in a non-crimp structure is based on the assumption that the crimp in the reinforcement makes elongation and reinforcement and resin will have lower initial modulus and poor energy distribution over a larger area in composites. These structures are specially engineered for composites applications so that they will perform in conjunction with resin and the synergy in composites could be enhanced. It is because they have good impact resistance (due to the absence of crimp), and when a force is applied on a specific area on them, they will distribute that force on the maximum area and can bear that force easily.

NCFs are different from other 3D and 2D structures due to the absence of crimp; this imparts the special characteristics in the structure such as the following:

1. The presence of crimp in any fabric structure causes low impact resistance as well as low energy dissipation. In NCFs, no crimp is present, so they provide high impact resistance as well as high

energy dissipation and are commonly used in that types of composites where these properties are necessary to be achieved [19].

2. Higher densities of fabrics can be achieved.
3. Better resin infusion because of the strong capillary action of straight yarns (in and out of plane) [20].
4. The NCF-reinforced composite materials have higher initial modulus [21].

In NCFs, layers of warp and weft are just stacked one over the other without any interlacement and these layers bind together by two separate warp yarns. For the weave design of NCF, the numbering of warp and weft yarns is done separately, as shown in Figure 5.23.

The weave design of six-layered non-crimped fabrics, as per the cross section shown in Figure 5.22, is given below:

	1	2	3	4	5	6	7
12		X	X	X	X	X	X
11		X	X	X	X		X
10		X	X	X			X
9		X	X				X
8		X					X
7							X
6	X	X	X	X	X	X	
5	X	X	X	X	X		
4	X	X	X	X			
3	X	X	X				
2	X	X					
1	X						

5.4 Unidirectional Fabric (2D NCF)

Unidirectional (UD) fabrics have superior mechanical properties than conventional fabrics only in one direction. The UD fabrics have all fibers/yarns oriented in one direction without any crimp. In other direction, there are no reasonable fibers/yarns. Ideally, UD fabrics have straight yarns that lie parallel to each other. Just laid yarns are not stable enough that could be handed. To give stability, yarns, for the ease of handling, are held together by weaving, warp knitting, stitching, or by a resin/gummy material. Woven (UD) fabrics have mostly plain weave. Ends/inch are kept much higher than picks/inch. Warp yarns are at higher tension and picks inserted are of very flexible yarn, which easily gets crimped. Mostly, the multifilament yarn of polypropylene or polyethylene is used in picks and 2–4 picks/inch are there. A fabric made in such a way is very fragile and is difficult to handle, so the

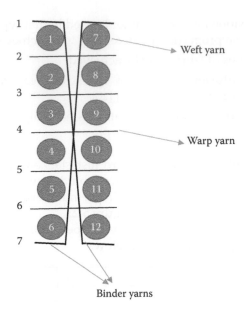

FIGURE 5.23
Cross section of six-layered non-crimped fabric.

fabric is passed through a hot press and thermoplastic weft melt and stick to fabric, thus giving a stable structure to handle.

Warp knitting machines used to make UD fabrics have straight picks of carbon/glass, inserted through air jet, and warp yarns are just to hold the picks from different points. Warps of warp knitting machines play only a negligible part in strength. One of the superiorities of a warp knitting machine-based structure is that multiaxial structure can be made easily.

5.5 Pile Fabrics

5.5.1 Terry Fabrics

A terry towel is known as a textile product having loop pile on one or both sides, generally covering the entire surface or forming strips, checks, or other patterns. A terry towel consists of different parts: end hem, cam, dobby, terry body, and selvages. It is not necessary that all parts will come in every towel.

Terry towel weaving machine uses two beams during weaving. One beam is used for base/ground fabric, while other beam is for the loop formation. Tension of both beams is kept different for both pile (slack) and ground (tight) beams. Generally, two-pick, three-pick, four-pick, five-pick, and

seven-pick terries are used for making terry towels. But most commonly, three-pick terry is produced. Three-pick terry means after each set of three picks insertion a full beat-up is made and one loop pile is formed on both sides of the fabric. The reed of the first two picks will be loose and that of the third pick reed will be fast in the three-pick terry design. The first pick holds one side of loop and the second pick holds other side of loop; after the third pick insertion, a full beat-up is made, which is responsible for the complete loop formation, as shown in Figure 5.24. Normally, four frames are used for making plain towel, two frames for pile, and two frames for ground.

Warps are ordered throughout the fabric width 1:1 or 2:2 piles and ground warps. In 1:1 warp order; each ground warp end is followed by a pile warp end, while in 2:2 warp order, each two ground warp ends are followed by two pile warp ends. The 2/1 warp rib weave design is used for both ground and pile yarns, but the only difference is that pile yarns are one pick ahead as compared to the ground yarns, as shown in Figure 5.25. Figure 5.25a shows the 1:1 drawing in reed plan, while Figure 5.25b shows the 2:2 drawing in reed plan of three-pick terry towel design.

By changing the arrangement of pile and ground yarns, a variety of designs could be achieved. On the basis of designs, towel can be classified into different types such as rice weave towel, tonal bird eye towel, mélange towel, one-sided terry towel, and two-sided terry towel.

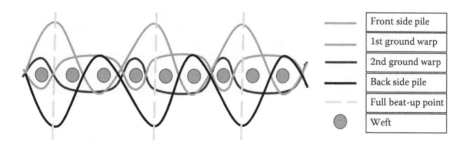

		Front side pile
		1st ground warp
		2nd ground warp
		Back side pile
		Full beat-up point
●		Weft

FIGURE 5.24
Full beat up and yarn arrangement in 3 pick terry.

(a)

3		X	X	
2	X			X
1	X	X		
	G	FP	G	BP

(b)

		X		
	X		X	X
	X	X		
	G	FP	BP	G

G = Ground warp, FP = Front pile, and BP = Back pile

FIGURE 5.25
1:1, 3 pick (a) and 2:2, 3 pick terry (b).

If any towel has end hem, dobby border design (other than plain), then we have to increase the number of frames depending upon the weave design. The dobby border design could be plain, twill, diagonal, satin, sateen, pin tuck, cord, diagonal cord, check pattern, stripes, arrow, weft inserts (satin), and diamond.

5.5.2 Carpet

Carpets have been known for luxury over centuries. The manufacturing is as old as that of fabrics. Both were produced manually, which required human skill and hard work.

Carpets offer a wide-ranging designs, colors, and textile fibers. The end use of the carpet decides the quality and type. The indoor use is different from that of outdoor rugs.

Construction, design and color, quality, performance, indoor climate properties (insulation, acoustics, safety), and environmental properties affect the final selection of carpet for specific end use.

Carpets offer comfort and warmth. Selection of a certain type depends on style/design, quality, comfort, sterility, cleanliness, climate, acoustics, energy, and human safety.

The main carpet fabric manufacturing methods involve tufting, weaving, and needle punching.

The tufting machine was invented by America in early 40s. This machine is different from the weaving machine. It is comparable to a big sewing machine. Yarn from creeled packages is fed to the machine and primary backing is also fed to it at the same time. Needles continually penetrate tuft into primary backing. The product is rolled as per the finishing requirement. It has the advantage of fast production and easy design change flexibility over weaving process. Following are important terms involved in tufted carpets.

Density is the determination of the number of tufts per inch across the width. It is space between two adjacent needle points. For example, 1/6 gauge means that six tuft rows per inch of a width.

Stiches per inch means a number of yarn tufts per running inch of a single tuft rows. Weight and density are subjective stiches.

Tuft height is measured from the surface of primary backing to the top of a tufted yarn. This affects the weight of the pile yarn. Tufted carpets involves the following:

1. Face yarn (tuft can be cut pile, loop pile, or a combination of both)
2. Primary backing fabric (woven or nonwoven fabric in which pile yarn is inserted)
3. Bonding compound (adhesive that fortifies yarn to primary backing)

4. Secondary backing (cushion added to provide extra stability to structure; this could be resistant to moisture and edge fraying made of textile or latex backings)

Many weaving methods are used by employing different types of looms. It involves two sets of yarn, that is, warp and weft are interlaced with each other.

Face yarn materials are natural carpet fibers like wool and cotton, whereas synthetic fibers used are nylon/polyamide (PA6, PA66), polypropylene (PP), acrylic, and polyester (PET). Their use depends upon the end use and desired engineered properties.

Textile yarns used are of two types: bulked continuous filament (BCF) and staple yarn.

Carpet face styles can be of different types: level loop pile, multilevel loop pile, multilevel cut pile, Saxony, Frieze, and cut and loop pile.

Woven carpets are made by three techniques, namely, Axminster, Velvet, and Wilton. These are differentiated on the basis of the weave design and binding of pile. These carpets have backs usually made of jute and cotton fiber. Polypropylene is also used. The yarn blend of 80/20 of wool/polyamide or 100% wool or synthetic yarn is used as well. Woven carpets made either with cut pile or with boucle present many options for colors and designs.

Needle punching, one of the nonwoven assembly techniques, is the third method used for carpet manufacturing. Desired layers of carded fibers are needle punched. These carpets are used as floor coverings with or without coated back. These are smooth and can be printed.

Specifications and testing standards involve the following:

1. Construction data

2. Measurements

3. Patterns

4. Colors

5. Shading

6. Pile reversal

Construction data is based on European EN 1307 standard for textile floorings. This includes a total weight of the carpet in g/m^2 ±15%. Pile weight indicates the yarn weight over the primary backing in g/m^2 +15/−10%. Pile height indicates the pile's height from the primary backing with ±1 mm allowance.

Measurements of standard lengths are based on the allowance of ±0.5%. Lengths supplied have edge of ±0%. Standard widths are supplied a verge of ±3 cm for 400- and 500-cm roll width. Margin for cuts (cut service) is +1/−0%.

For area rugs, the length is supplied with ±2% and width with an allowance of ±1.25%. Modular (segmental) lengths and widths have ±0.2% range within the same batch.

References

1. R. Mark and A. Robinson, *Principles of Weaving*, Manchester, UK: The Textile Institute, 1976.
2. Y. Liu, J. Zhu, Z. Chen, Y. Jiang, C. Li, B. Li, L. Lin, T. Guan, and Z. Chen, Mechanical properties and microstructure of 2.5D (shallow straight-joint) quartz fibers-reinforced silica composites by silicasol-infiltration-sintering, *Ceram. Int.*, 38, 795–800, 2012.
3. Y. Z. Wan, G. Zak, S. Naumann, S. Redekop, I. Slywynska, and Y. Jiang, Study of 2.5-D glass-fabric-reinforced light-curable resin composites for orthotic applications, *Compos. Sci. Technol.*, 67(13), 2739–2746, 2007.
4. J. Jekabsons and J. Varna, Micromechanics of damage accumulation in a 2.5D woven C-fiber/SiC ceramic composite, *Mech. Compos. Mater.* 37(4), 289–298, 2001.
5. A. Hallal, R. Younes, F. Fardoun, and S. Nehme, Improved analytical model to predict the effective elastic properties of 2.5D interlock woven fabrics composite, *Compos. Struct.*, 94(10), 3009–3028, 2012.
6. A. Hallal, R. Younes, S. Nehme, and F. Fardoun, A corrective function for the estimation of the longitudinal Young's modulus in a developed analytical model for 2.5D woven composites, *J. Compos. Mater.*, 45(17), 1793–1804, 2011.
7. S. Nauman and I. Cristian, Geometrical modelling of orthogonal/layer-to-layer woven interlock carbon reinforcement, *J. Text. Inst.*, 106(7), 725–735, 2015.
8. F. Boussu, M. Lefebvre, D. Coutellier, and D. Vallee, Experimental and high velocity impact studies on hybrid armor using metallic and 3D textile composites, in *The Composites and Advanced Materials Expo*, October 13–16, 2014, Orlando, Florida.
9. F. Boussu, I. Cristian, and S. Nauman, General definition of 3D warp interlock fabric architecture, *Compos. Part B*, 81, 171–188, 2015.
10. D. Sun and X. Chen, Three-dimensional textiles for protective clothing, in X. Chen (ed.), *Advances in 3D Textiles*, London, UK: Elsevier, 2015.
11. H. Gu and Z. Zhili, Tensile behavior of 3D woven composites by using different fabric structures, *Mater. Des.*, 23(7), 671–674, 2002.
12. L. W. Taylor and L. J. Tsai, An overview on fabrication of three-dimensional woven textile preforms for composites, *Text. Res. J.*, 81(9), 932–944, 2011.
13. Z. J. Grosicki, *Watson's Textile Design and Colour*, 7th ed., London, UK: Butter Worths Group, 1975.
14. S. Alderman, *Mastering Weave Structure*, Loveland, USA: Interweave Press, 2004.
15. U. Riedel and J. Nickel, Applications of natural fiber composites for constructive parts in aerospace, automobiles, and other areas, *Biopolym. Online*, 2005. DOI: 10.1002/3527600035.bpola001.

16. M. Umair, Y. Nawab, and M. H. Malik, Development and characterization of three-dimensional woven-shaped preforms and their associated composites, *J. Reinf. Plast. Compos.*, 34(24), 2018–2028, 2015.
17. A. Mountasir, G. Hoffmann, and C. Cherif, Development of weaving technology for manufacturing three-dimensional spacer fabrics with high-performance yarns for thermoplastic composite applications: An analysis of two-dimensional mechanical properties, *Text. Res. J.*, 81(13), 1354–1366, 2011.
18. A. E. Bogdanovich, Advancements in manufacturing and applications of 3D woven preforms and composites, *16th International Conference on Composite Materials*, 2006, pp. 1–10.
19. F. Edgren, D. Mattsson, L. E. Asp, and J. Varna, Formation of damage and its effects on non-crimp fabric reinforced composites loaded in tension, *Compos. Sci. Technol.*, 64(5), 675–692, 2004.
20. J. Verrey, V. Michaud, and J. A. E. Månson, Dynamic capillary effects in liquid composite moulding with non-crimp fabrics, *Compos. Part A Appl. Sci. Manuf.*, 37(1), 92–102, 2006.
21. D. S. Ivanov, S. V. Lomov, A. E. Bogdanovich, M. Karahan, and I. Verpoest, A comparative study of tensile properties of non-crimp 3D orthogonal weave and multi-layer plain weave E-glass composites. Part 2: Comprehensive experimental results, *Compos. Part A Appl. Sci. Manuf.*, 40(8), 1144–1157, 2009.

6

Textile Structures for Jacquard Weaving

Danish Mahmood Baitab and Adeela Nasreen

CONTENTS

6.1 Jacquard Shedding

Jacquard shedding is used to produce figured woven fabrics used for decorative purposes. Design in fabric during weaving depends on the shedding of harnesses. A more complicated design requires a more number of harnesses. Dobbies are limited to 32 harness frames repeat of woven fabric, but when fabric design requires more warp repeats, jacquard shedding is employed. Also for ornamental intricate design of warp yarns are required to control threads individually, so jacquard is a must for such designs. There are no harness frames used on jacquards, but warps are individually raised, lowered, or kept at their position as required in each repeat. Jacquard is used to

produce fancy fabric like brocades, damasks, extra-warp or extra-weft figured fabrics, double cloths, swivel fabrics, leno brocades, tapestries, etc. [1].

6.1.1 Jacquard Machine Parts

Jacquard machine parts are as follows:

1. Pattern chain
2. Motor
3. Pattern cylinder
4. Needle
5. Knife
6. Harness cord
7. Neck cord and comber board
8. Top board
9. Hook
10. Grid bar
11. Dead weight
12. Spring board
13. Needle board [1]

6.2 Classification of Jacquard

Jacquards are of two types:

- Mechanical
- Electrical

6.2.1 Mechanical Jacquard

In mechanical jacquard, design is mechanically fed to the jacquard in the form of punch cards. Design cards are punched on paper cards according to the design of fabric. For one pick of design, one card is punched. The total number of punch cards will be equal to the total number of picks in design. Generally, three types of mechanical jacquards are used [1]:

- Single lift and single cylinder (SLSC)
- Double lift and single cylinder (DLSC)
- Double lift and double cylinder (DLDC)

6.2.1.1 Single Lift and Single Cylinder

Figure 6.1 shows the simplified side view of SLSC jacquard. If the machine has the capacity to handle 300 ends independently, then it requires 300 hooks (one per end) that are vertically arranged and 300 needles (one per hook) that are horizontally arranged. For example, the needles can be arranged in 6 rows and each row will have 50 needles. In the side view, only six needles (one per horizontal row) are visible. Hooks, which are connected to individual ends through nylon cord (harness), are also arranged in 6 rows, and each row is having 50 hooks. One knife is responsible for controlling the movement (lifting and lowering) of one row of hooks. However, whether a hook will be lifted or not will be ascertained by the selection mechanism which is basically a punched card system mounted on a revolving cylinder having square or hexagonal cross section. The needles are connected with springs at the opposite side of cylinder. Therefore, the needles always exert some pressure in the right-hand side as shown in Figure 6.1. So, if there is a hole in the punch card corresponding to the position of a needle, then the needle will be able to pass through the hole, and thus the needle will remain in upright position, thus making it accessible to the knife when the latter has started its upward movement after descending to the lowest height. On the other hand, if there is no hole, then the needle will be pressed toward the left side against

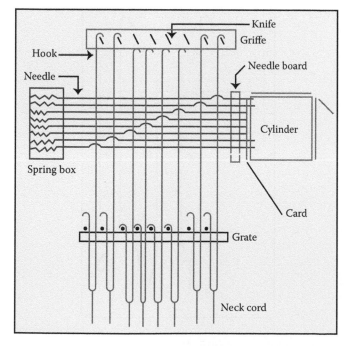

FIGURE 6.1
Single lift and single cylinder.

the spring pressure. Thus, the kink (which partially circumscribes the stem of a hook) present in the needle presses the hook toward the left side making the latter tilted enough from the vertical plane so that the knife misses it while moving upward. Therefore, the presence of a hole implies selection (ends up) and vice versa. A hole in this case is tantamount with a peg used on the lag of dobby shedding system.

In the case of SLSC jacquard, if the loom speed is 300 picks/min, the cylinder will turn 300 times/min (5 times/s) and the knives should also reciprocate (up and down) 300 times/min, thus limiting the loom speed. When a particular hook (and the corresponding end) has to be in up position in two consecutive picks, in between the two peaks, it descends to its lowest possible height (determined by the grate) and then moves up again. Thus, it produces bottom-closed shed. This happens as one end is controlled by a single hook.

6.2.1.2 Double Lift and Single Cylinder

The double lift machine forms a semi-open shed. A schematic view of DLSC jacquard is shown in Figure 6.2. It remains open for a longer time than the closed one, thus increasing the speed to about 160 picks/min. At this speed,

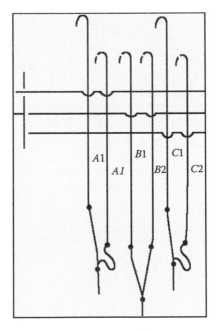

FIGURE 6.2
Double lift and single cylinder.

each knife moves up or down at the rate of 80 times/min, while the cylinder moves 160 times/min. The rising and falling sheds act as counterpoise to each other, and therefore less power is required for the formation of a shed. As there is less strain on the warp, relatively weaker yarn can be tolerated for working on the double lift machine. As the weft is beaten up in the *crossed shed* in the double lift machine, the weft cannot recede from the fell of the cloth and thus produces a better cover of distribution of threads. In a single lift machine, the weft is beaten up in the *closed shed* in which the pick has a tendency to slip back from the fell of the cloth. This also enables one to insert more picks per unit space than in the case of a single lift machine. If the same warp threads are required to be raised on two consecutive picks, one of the hooks lifts the threads to the top shed line. When that hook begins to come down, the threads controlled by that hook also begin to descend with it. At the same time, another card is against the needle board with a hole punched for the same needle. Therefore, the paired hook starts ascending with its griffe.

The ascending hook and descending hook therefore meet half way between their upper and lower limits of movement. The descending threads are thus raised again by the ascending hooks to form the top shed line. The bottom shed line remains stationary. Threads that are required to be down from the previous raised position continue their movement to the bottom shed line. In this way, semi open shed is formed.

6.2.1.3 Double Lift and Double Cylinder

In DLSC jacquard, the limiting factor for the increase in speed is the movement of the cylinder. A schematic view of DLDC jacquard is shown in Figure 6.3. In order to obviate this difficulty, two cylinders are provided on this machine. With such an arrangement, the two cylinders present their cards at alternate picks. The loom could be operated at a speed of 180 picks/min. Each cylinder controls alternative picks. There are two hooks attached to one neck cord, both of these hooks are controlled by two needles, and these needles and hooks are placed opposite to each other. The knives of one griff are also angled in the direction opposite to that of the other set of knives. One set of knives rises as the other is lowered. The rising griff engages the hooks selected by the card, which is presented by the cylinder for that pick. It is therefore necessary that half of the cards that say all odd-numbered cards such as 1, 3, 5, etc. are laced together to form an endless chain for one cylinder and the remainder, all even-numbered cards such as 2, 4, 6, etc. are laced for the other cylinder. If the odd-numbered cards are laced forward, then the even-numbered cards are laced backward. This is necessary because one cylinder turns in clockwise direction, while the other turns in an anticlockwise direction. A peculiar lacing of the odd- and even-numbered cards separately is a special feature of a DLDC jacquard.

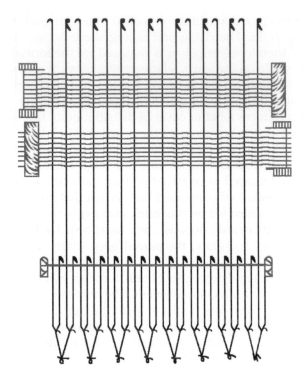

FIGURE 6.3
Double lift and double cylinder.

6.3 Open Shed Jacquard

If the blank portion of the card laces the needle, the ends are kept down in the bottom shed line. If a hole is against a needle, one of the hooks will be lifted by one griffe. If the same end is required to be up for the next pick in succession, the other hook moves up with the other griff so that the slack cord of the descending hook is taken up by the ascending hook and the neck cord remains in the unaffected raised position as a sequence of hooks raising in open shed jacquard shown in Figure 6.4.

6.4 Electronic/Electrical Jacquard

In Figure 6.5, module of electronic jacquard is shown. Labels mentioned for elements of electronic jacquard module are explained in Table 6.1.

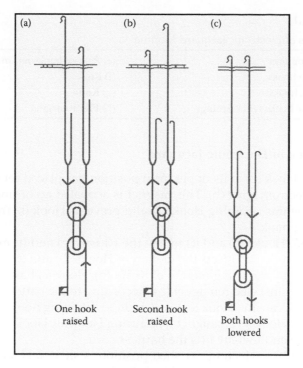

FIGURE 6.4
Open shed jacquard.

FIGURE 6.5
Module of electronic jacquard.

TABLE 6.1

Elements of Electronic Jacquard Module

(a) Hoist pulleys	(e) Release hooks return springs
(b) Mobile hooks	(f) Knife
(c) Mobile hooks	(g) Knife
(d) Release hooks return springs	(h) Electromagnet

6.4.1 Working of Electronic Jacquard

In Figure 6.6, Hook (b) in its upper most position has placed retaining Hook (d) against electromagnet (h). This magnet is activated according to the pattern, briefly retains retaining Hook (d), and prevents Hook (b) from hooking onto retaining hook.

In Figure 6.7, Hooks (b) and (c) follow the Knives (g) and (f) moving up or down. Double roller (a) offsets the motion of Hooks (b) and (c).

In Figure 6.8 by the rising motion of Knife (g), Hook (c) has placed retaining Hook (e) against electromagnet (h). According to the pattern, the magnet is not activated, causing Hook (c) to catch on to retaining hook.

In Figure 6.9, Hook (e) is caught on retaining Hook (e). Hook (b) follows the rising Knife (f) and thereby lifts the harness cord.

In Figure 6.10, in which upper shed position is shown, Hook (c) remains hooked onto retaining Hook (e). Hook (b) has placed retaining Hook (d)

FIGURE 6.6
Position 1.

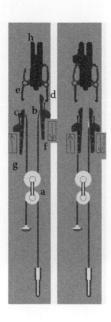

FIGURE 6.7
Position 2.

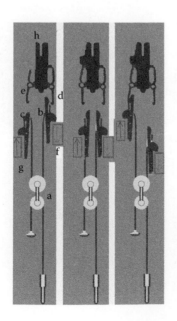

FIGURE 6.8
Position 3.

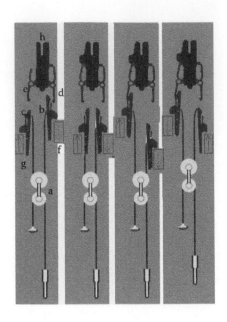

FIGURE 6.9
Position 4.

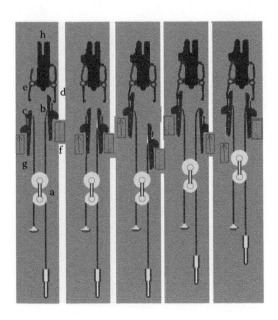

FIGURE 6.10
Position 5.

FIGURE 6.11
Position 6.

against electromagnet (h) by the rising motion of Knife (f). According to the pattern, the magnet is not activated, causing Hook (b) to be held by retaining hook.

In Figure 6.11, Hooks (b) and (c) remain held by retaining Hooks (d) and (e). Knives (g) and (f) are in rising and descending motion, respectively [2].

6.5 Computer-Aided Designing for Jacquard Structures

6.5.1 Introduction

Jacquard designing is for those patterns having repeat on ends greater than 32 frames. Contrary to the tappet and dobby shedding, the jacquard system has harness or hook responsible for the shedding. Hooks are used to raise/lower the individual yarns. Therefore, designs having unlimited repeat pattern like flowers, logos, text, etc. can be produced using jacquard.

There are no heddle frames used on jacquard, but ends are raised or lowered individually as per requirement in each repeat. Jacquard machine is used to produce fancy fabrics like damask, leno, or brocades fabrics. There are two main types of jacquard as already discussed in jacquard shedding portion. These are mechanical jacquard and electronic jacquard.

6.5.2 How to Feed Design on Mechanical Jacquard

To make design using mechanical jacquard, pattern cards that are punched and laced together according to repeat are used. Circular holes are punched in the card corresponding to needles. A hole in card is responsible to raise warp end and make top shed line, and blank is responsible to lower the warp end and make bottom shed line. Pattern card has design holes for needles and peg holes to fix card in exact position and lace holes for joining cards together as you can see fully punched card in Figure 6.12.

One card is used for one pick only, so there are as many cards as many picks in the repeat of design. Comber board made of close-grained wood is a long perforated board extending the width of loom, and the objective of comber board is to spread harness cords uniformly, which determines the warp density in the cloth. The reed number must correspond to the number of holes per unit length in the comber board. The number of holes in width of comber board is equal to the number of needles in the short row of jacquard. The number of holes in lengthwise direction of comber board depends on the warp density of cloth being made.

For example, if jacquard machine contains eight needles in the short row, there will be eight holes in the short row of comber board. While if reed contains 96 ends/inch, there will be 12 holes in each inch of long row, thus giving in total 96 holes/inch in the comber board lengthwise. Once a comber board is threaded and drilled, the width of fabric and the width of repeat are calculated. The thread per unit width could be decreased but never increased.

In tappet and dobby shedding, the number of ends per centimeter can vary. With jacquard shedding, the number of ends per centimeter is determined by harness cords. There is no possibility of weaving a fabric with more ends

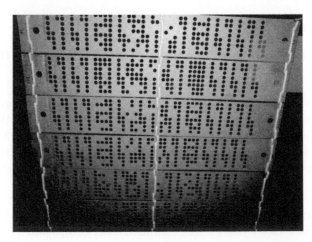

FIGURE 6.12
Fully punched pattern card.

per centimeter than the harness has. However, it is possible to weave design with a less number of ends per centimeter by using casting out [3].

6.5.3 Casting Out

Casting out means to leave selected hooks and harness cords so that they become idle. For example, the maximum hooks are 400 and the maximum warp density is 32 ends/cm. So, the size of 400 thread repeats will be 400/32 equals to 12.5 cm. If we want to produce 24 ends/cm, the maximum required hooks will be 24 × 12.5 equals to 300 hooks. We have to leave 100 hooks idle, so each fourth row will be idle or we can make every fourth hook idle [1].

6.5.4 How to Feed Design on Electronic Jacquard Loom

For making fabric on electronic jacquard machine, different designing software are used like Scot weave (for fabric) and Eat software (for terry towel) on which from image processing, resolution setting, ends and picks calculation, and weave assigning method to cast out all process are done, and then this information is transferred to floppy disk or USB and fed on loom.

For computer-aided designing of jacquard, certain kinds of computer software are used, for example, Scot weave. For jacquard production, in scot weave image (design) is first converted to loom readable form. For that purpose, certain steps are carried out as follows:

1. Artworks
2. Jacquard designer
3. Jacquard loom

6.5.4.1 Artwork

Artwork is the initial step of jacquard process. In artwork, image is imported. Every image has too many colors (dominant and not readable). For jacquard image, colors are reduced to two colors. Reducing image color is also known as color compression. Artwork is also responsible for adjusting the size of the design or image in terms of ends/inch and picks/picks. Design tracing can also be done in artwork. Moreover, editing of design, copy paste of design, and mirroring of design are done in artwork.

6.5.4.2 Jacquard Designer

After working on design in artwork, it is imported to jacquard designer. In jacquard designer, different weaves are applied to different colors. In other words, for each different color, a different weave is applied. Scot weave has a treasure of different kinds of built-in weaves and counts of yarns. After

applying weaves, yarns of required count are applied in warp and weft to see the design whether it is correct or not.

6.5.4.3 Jacquard Looms

After applying weave and yarns to the design, then it is checked. If the design is correct, then it is imported to jacquard window in Scot weave. Here, harnesses (hooks) are applied to each yarn. Jacquard loom window is also responsible for adjusting the selvedge of the fabric to be produced and also specific yarn to be run for the specific weave in weft. After that, design is saved in a floppy disk and transferred from computer software to jacquard loom [3].

References

1. S. Adanur, *Hand Book of Weaving*. Alabama, USA: Auburn University, 2001.
2. Jacquard Shedding – Jacquard Shedding Mechanism – Textile Learner [Online]. Available: http://textilelearner.blogspot.com/2012/02/jacquard-shedding-jacquard-shedding.html (Accessed 08 March 2016).
3. R. Mark and A. Robinson, *Principles of Weaving*. Manchester, UK: The Textile Institute, 1976.

7

Color and Weave Effect

Muhammad Zohaib Fazal and Muhammad Imran Khan

CONTENTS

7.1 Introduction

Conventional textiles are mostly used for esthetic and ornamental purposes. Color and luster are the two important aspects that enhance the esthetic appeal of a fabric. While purchasing a garment, a consumer is attracted firstly by the color and then by the luster of the fabric. Both color and luster are perceived because of the reflection of light from the surface of the textile fabric. The reflection of light from the surface is dependent on the yarn hairiness, weave design (float), and applied finishes. In this chapter, the relationship between color and weave design is discussed in detail.

7.2 Color

Color is the visual perceptual property corresponding in humans to the categories called red, blue, yellow, etc. Visual perception is the ability to interpret the surrounding environment by processing information that is contained in the visible light. The resulting perception is also known as eyesight, sight, or vision. Color derives from the spectrum of light interacting in the eye with the spectral sensitivities of the light receptor cells.

Color categories and physical specifications of color are associated with three factors: light source, object, and the observer. Light source is the incident light falling on the object. A portion of incident light is absorbed by the object and remaining is reflected to the observer (human eye). All the three factors affect the interpretation of colors.

7.2.1 Color and Light Theory

Vision is due to the stimulation of the optic nerves (light receptor cells) present in human eye by light either directly from its source or indirectly after reflection from other objects. With reference to light emission, all objects, either natural or artificial, can broadly be divided into two categories: luminous objects that emit light, for example, sun, stars, gas, electric light, etc. and nonluminous or illuminated objects that do not emit their own light.

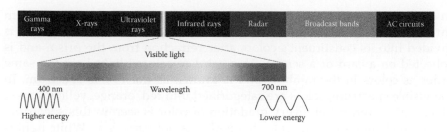

FIGURE 7.1
Electromagnetic spectrum.

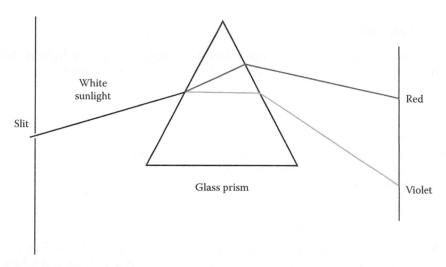

FIGURE 7.2
Newton's experiment: Sunlight is the source of all colors.

Illuminated objects are visible because of their property of reflecting light. These objects do not only reflect light but also transmit some portion of it.

Electromagnetic radiation is the radiant energy released by certain electromagnetic and gravitational processes. Light is also a form of energy and is a part of the electromagnetic spectrum, as shown in Figure 7.1. Humans are capable of visually sensing a narrow range of this electromagnetic spectrum, that is, 400–700 nm wavelength (Bohren and Clothiaux, 2008), which is known as visible span or visible spectrum. Shorter wavelength radiations have a higher frequency and higher energy; as the wavelength increases, both the frequency and energy decrease.

A series of experiments were performed by Sir Isaac Newton during 1666–1672, enabling us to understand the relationship between light and color (Swenson, 2010). Newton's simple experiment is shown in Figure 7.2, and this experiment determines the composition of white light as it splits into its colors and hence demonstrates that light is the source of all colors.

In this experiment, a beam of white light is allowed to enter a dark room through a small slit. When this light beam passes through a prism, it is divided into its constituent's colors after refraction from the prism and is collected on a card or a screen. This band of colored light is in the same order as colors in the rainbow and is known as the visible spectrum. In the visible spectrum, colors are categorized into red, orange, yellow, green, blue, indigo, and violet. Every gradation of color is seen in this spectrum, although change from one color to another is not very clear. White light is pure light and it consists of all the natural and artificial colors. Any color can be described by its wavelength.

7.2.2 Perception of Color

Newton's phrase "the rays are not colored," makes it clear that light is not intrinsically colored but it is the sensation that light rays produce in the human brain. Color exists only in mind. Every human being perceives color according to his or her own visual perception. It is a perceptual response to light that enters the eye either directly from self-luminous light sources or indirectly from light reflected by illuminated objects. Colors cannot be described without studying the properties of the human visual system. Since color is the sensation produced by light rays on the retina of the human eye and perceived as a specific color by the brain.

7.2.3 Human Visual System

Human eye is the organ that receives light and is responsible for vision. It is made up of three layers: an outer most layer composed of cornea and sclera, a middle layer consisting of choroid and iris, and an inner and the most important layer consisting of retina. Retina contains two types of light receptor cells: rods and cones. The rod cells are more numerous, around 90 million, and are more sensitive to light than the cones. Rods are of only one type. They are insensitive to color and responsible for night vision (capability to see in low-light conditions).

A total of 6–7 million cones provide the eye's color sensitivity and function best in relatively bright light, as opposed to rod cells, which work better in dim light. Cones are much concentrated in the central yellow spot called macula. In the center of macula is the "fovea," a 0.3-mm-diameter rod-free area with very thin, densely packed cones, enabling humans to distinguish between different tint and shade of same color. They are able to perceive finer detail and more rapid changes in images, because their response time to stimuli are faster than those of rods. Cones are less sensitive to light but allow color perception due to their three types of light receptor cells, namely:

- S-cones (blue) respond to light of short wavelengths with peak at 420 nm

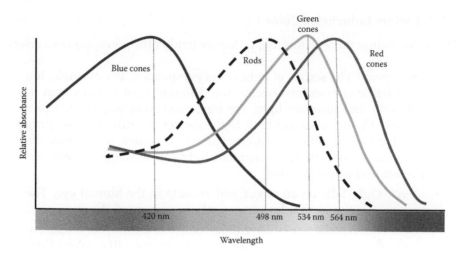

FIGURE 7.3
The relative sensitivity of cone cells.

- M-cones (green) respond to light of medium wavelengths with peak at 530 nm
- L-cones (red) respond to light of long wavelengths with peak at 560 nm

Figure 7.3 illustrates the relative sensitivity of cone cells. These cone sensitivities support the trichromatic theory, as all colors of the visible spectrum can be seen by mixing the three primary colors (red, green, and blue). The human eye can distinguish about 10 million colors (Judd and Wyszecki, 1963).

Color blindness is the inability to identify and distinguish certain colors. It can be due to the presence of more than three types of cones, giving color blind patients tetrachromatic vision. Color blind patients can also suffer due to absence of some type of cones giving range of defective visions.

7.2.4 Properties of Object

The color of an object is not the intrinsic property of the matter of which object is made. All illuminated objects have the property to break down the falling light into its constituent rays of different wavelengths. Then these objects selectively absorb, transmit and reflect the portion of these rays. So, the color, which humans perceive is due to the reflected rays of specific wavelength from the object. Color of an object is determined by the character and intensity of the light rays reflected by it. White color is due to the total reflection and black color is due to the total absorption of incident light falling on an object. Gray materials reflect and absorb almost equal quantities of incident light.

7.2.5 Factors Influencing Color

There are mainly three factors that influence or alter the perception of a color.

- *Light source*: The source of light is very important. For example, if a dyed fabric is examined under two different light sources, that is, sun light and tungsten light, the perceived color will be different because the intensities of both light sources are different. Keeping the object and observer same, change in light source, will change the perceived color. Change in incident light cause change in reflected portion and hence the color.
- *Object*: Light falls on an object and reflects to the human eye. The precision of the same color varies with the change of object.
- *Human eye*: It is the sensory organ to receive and interpret the sensations accordingly. If the eye has any discrepancy like color blindness, then there is an insensitivity to particular colors, which in turn can affect the color perception.

7.2.6 Basic Terminologies Related to Color

There are many terms used to describe colors; some of the common terms are as follows:

1. *Hue*: Hue is somewhat synonymous to what we usually refer to as "color." It is an attribute of a color by which it is discernible as red, green, blue, etc. Different hues have different wavelengths in the spectrum. It is independent of intensity or lightness.
2. *Saturation*: It is also called the intensity of color and is used to express the strength or purity of color. Saturation is the measurement of how different the color is from pure gray, as shown in Figure 7.4. When a hue is strong and bright, it is said to be high in intensity, and when a color is faint, dull, and gray, it is said to be low in intensity.
3. *Value*: Value refers to the amount of luminosity in the color. It is measured by the amount of light in a color. Value is the lightness or darkness of the color. Different values can be obtained by adding black or white to a color.
    a. *Tint*: Tint of a color (hue) is made by adding white so that it becomes lighter. For example, pink is a tint of red.
    b. *Shade*: Shade is made by adding black to a color (hue) so that it becomes darker. Maroon is a shade of red. Figure 7.5 depicts the tint and shade of red.
    c. *Tones*: By adding gray to a pure color (hue), tones of a color are obtained.

FIGURE 7.4
Intensity of green color.

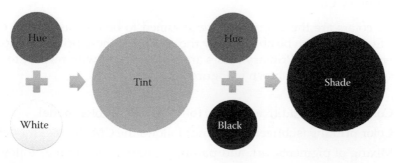

FIGURE 7.5
Tint and shade of red color.

4. *Harmony*: When two or more colors blend well, color harmony is produced. A combination of hues that exist in harmony is pleasing to the eye.

5. *Contrast*: Contrast is the noticeable level of difference between two colors. When two or more colors oppose each other, or appear dissimilar, a contrast is created.

7.2.7 Color Mixing

There are two theories regarding color mixing. These theories of color mixing are based on the concepts of reflection and absorption.

1. *Light theory (additive mixing)*: When lights of different colors fall on a body and are reflected back, the resultant reflected ray that falls on the retina of the eye is the sum of all the individual reflected rays, which is called additive mixing. For example, in computer monitors, television screens, and in overlapping projected colored lights that are used in theatrical lighting, the phenomena of additive color mixing occur.

2. *Pigment theory (subtractive mixing)*: The mixing of dyes, inks, paint pigments, or natural colorants follows the idea of absorption of light. When different pigments or dyes are mixed together, their individual absorptions sum up in the resultant mixture, thus increasing the amount of absorption of light falling on them and, in the result, decreasing the amount of reflected light. Hence, it is called subtractive mixing. A printer combining several different colored inks on the top of each other on a paper and transparent photographic layers that are sandwiched together to produce color images are examples of subtractive mixing.

7.2.8 Primary Colors

Primary colors are the colors that are assumed to be the basis of all other colors. Every color can be obtained by the mixing of these primary colors in different proportions. Primary colors are also different with respect to light and pigment theory, as shown in Figure 7.6.

- Colored lights (additive mixing) follow the RGB color model.
- Color printing (subtractive mixing) follows the CMYK color model.
- Mixing of pigments, art, and painting (subtractive mixing) follows the RYB color model.

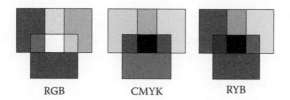

RGB CMYK RYB

FIGURE 7.6
Primary colors.

7.2.9 Color Wheel

The color wheel shows relationships between colors. The color wheel, as shown in Figure 7.7, is divided into three color areas: primary, secondary, and tertiary colors.

- Primary colors are on equal distance away from each other on the color wheel. These colors are used to make all other colors.
- Secondary colors are a mixture of two primary colors. They are on the color wheel between each of the primary colors and are on equal distance away from each other.
- Tertiary colors are a mixture of a primary color and a secondary color.

FIGURE 7.7
Color wheel.

7.2.9.1 Complementary Colors

Complementary colors are opposite, that is, at 180 degree angle on color wheel. These colors provide the most visual contrast when used together.

7.2.9.1.1 *Effect of Fatigued Nerves*

When a color is looked at, the corresponding nerves are excited. If the gaze is continued for a considerable time, the nerves become fatigued, while the other nerves are resting. When gaze is transferred to another surface, the rested nerves produce sympathetically an after-image, which is complementary in color to the first color. The exhaustion of the color nerves can cause a color to appear duller or even change the appearance of a color. If someone looks at red for certain time, blue appears greener; yellow appears greener; orange appears yellower; and green appears bluer. This phenomenon causes problems in examining dyed cloths. We should switch from one color to another, to avoid color disturbance.

7.2.10 Psychological Effects of Colors on Human

Colors have psychological effects on human beings. Colors refer to emotions. A more responsive behavior of a person toward colors indicates a greater affectivity by colors. Colors can also affect human moods and behaviors, causes to alter decisions or point of view. These effects can be either pleasant or nasty. There exists an equivalency between colors and feelings. Reactions to colors can also be used as diagnostic tools for some therapies. There exist different theories of color psychology, a vast and complex field beyond the scope of our subject.

7.2.10.1 Warm and Cool Colors

Colors on the color wheel, as shown in Figure 7.8, are classified into two categories depending upon their effects.

> *Warm colors* make objects seem warmer, closer, and cozier than they actually are. Warm colors, such as yellow and red, tend to advance and make the walls seem closer. They are good for large, uninviting rooms that one wants to make cozier and welcoming. Warm and bright colors give the illusion of being closer to a viewer within a composition.

> *Cool colors* make objects feel cool and roomy. They are often used in rooms that get a lot of sunlight so that they do not feel as hot. Cool colors, such as green and blue, tend to recede and make the walls seem farther away. This makes them a good choice for small, narrow rooms that one wants to make them seem more spacious. Cool and dull colors appear to be farther away. Advancing and receding of colors are shown in Figure 7.9.

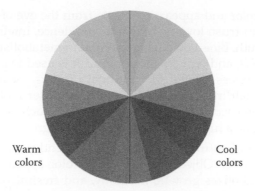

Warm
colors

Cool
colors

FIGURE 7.8
Warm and cool colors.

FIGURE 7.9
Advancing and receding colors.

7.2.10.2 Neutral Colors

White, black, and gray are considered to be neutral colors.

7.2.10.3 Effect of Colors

Every color imprints its effects on its observers. Living in an environment of a specific color, the inhabitants will behave and act accordingly.

Red color is brilliant, high energy, powerful, and cheerful; gives the impression of warmth; and appears to advance slightly toward the observer. It enhances the human metabolism, causes a raise in blood pressure, and aids in digestion. Red color is highly visible and due to which it is used in stop or danger signs and fire equipments. Red is also known to be a color of love and happiness.

Blue is a cold color and appears to recede from the eye of the observer. It symbolizes trust, loyalty, wisdom, confidence, intelligence, calmness, and truth. Blue slowdowns the human metabolism, produces a calming effect, and reduces the appetite. It is used to promote products and services related to cleanliness (filters, cleaning liquids), air and sky (airlines, air conditioners), and water and sea (mineral water). When used together with warm colors such as yellow or red, blue can create a high-impact and vibrant design.

Green is a color of nature. It is retiring and slightly cold color, having a soothing effect. Objects having green color appear cheerful and fresh. It symbolizes growth, harmony, and freshness. Green has a strong emotional correspondence with safety.

Yellow is very luminous, vivid, and bright color. It conveys the idea of purity, warmth, boldness, and happiness. It encourages activity, communication, and memory recall. Yellow appears more distinctly to advance to the eye.

Orange is a very strong color. It possesses warmth and brightness and encourages appetite. Orange symbolizes friendship, boldness, and socialness.

Purple combines the stability of blue and the energy of red. Purple is the color of royals. It encourages spirituality, imagination, and relaxation. It is associated with wisdom, dignity, independence, creativity, mystery, dreams, and magic.

7.2.11 Color Schemes

Colors when used in combinations may be used in different possible arrangements depending on the requirement. They are used to create style and appeal. Colors that create an esthetic feeling when used together will commonly accompany each other in color schemes.

7.2.11.1 Monochromatic Color Scheme

In this type of scheme (Figure 7.10), only one color or hue is used but with different strengths and textures (tints, tones, and shades) to make it more interesting. As a result, the energy is more subtle and peaceful in this scheme due to a lack of contrast of hue. It appears to be unified, harmonious, and professional.

7.2.11.2 Achromatic Color Scheme

Colors that lack chromatic content (hue) are said to be achromatic (without color), unsaturated, or near neutral, as shown in Figure 7.11. Pure achromatic colors include black, white, and all grays, while near neutrals include browns, tans, pastels, and darker colors. Near neutrals can be of any hue or lightness.

FIGURE 7.10
Monochromatic color scheme of red color.

FIGURE 7.11
Achromatic color scheme.

Neutrals are obtained by mixing pure colors with white, black, or gray, or by mixing two complementary colors. Black and white combine well with almost any other color. Black decreases the apparent saturation or brightness of colors paired with it, and white shows off all hues to equal effect.

7.2.11.3 Complementary Color Scheme

Complementary colors are directly across from each other on the color wheel as shown in Figure 7.12. For the best use, desaturate the cool colors rather than the warm ones. The high contrast of complementary colors creates a vibrant look, especially when used at full saturation. This color scheme must be managed well so it is not jarring. Complementary color schemes are tricky to use in large doses, but work well when something is to stand out or to be highlighted.

7.2.11.4 Split-Compliment Color Scheme

A split-compliment color scheme is a variation of the complementary color scheme. It includes a main color and the two colors on each side of its complementary (opposite) color on the color wheel shown in Figure 7.13. This

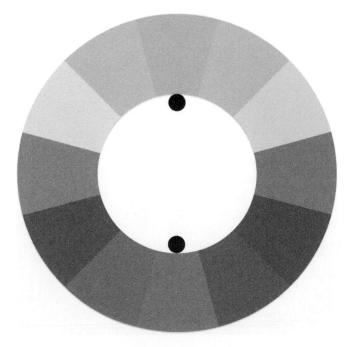

FIGURE 7.12
Complementary color scheme.

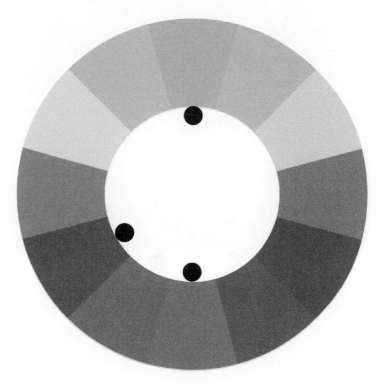

FIGURE 7.13
Split-compliment color scheme.

color scheme has the same strong visual contrast as the complementary color scheme, but has less tension. An example of a split-compliment color scheme could be green, violet–red, and red–orange.

7.2.11.5 Double Split-Complement Scheme

A double split-complement scheme uses two pairs of complements, one space apart on the color wheel. An example is red, green, orange, and blue.

7.2.11.6 Analogous Color Scheme

These colors are located next to each other on the color wheel, such as blue, blue–green, green, red, red–orange, and orange as shown in Figure 7.14. One color is chosen to dominate, a second to support, and the third color is used (along with black, white, or gray) as an accent. Analogous colors are sometimes called harmonious colors. They usually match well and create serene and comfortable designs. Analogous color schemes are often found in nature and are harmonious and pleasing to the eye.

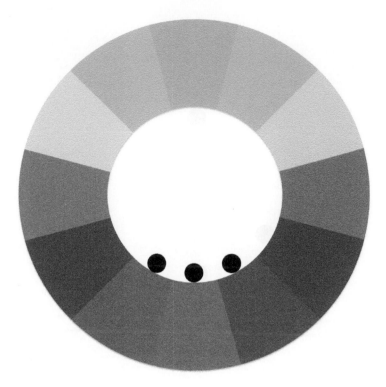

FIGURE 7.14
Analogous color scheme.

7.2.11.7 Triadic Color Scheme

It uses colors at the points of an equilateral triangle (three colors spaced equally on the color wheel). These are sometimes called balanced colors, as shown in Figure 7.15. Examples include red, blue, and yellow; and green, orange, and purple. Triadic color schemes tend to be quite vibrant, even if pale or unsaturated versions of hues are used. To use a triadic harmony successfully, the colors should be carefully balanced, one color dominating and the two others for accent.

7.3 Color and Fabric

Color can be imparted to a textile fabric by several methods, starting from the fiber, to sliver, roving, yarn, and fabric. Color in the final or end product will mainly depend upon the stage and process at which the textile product is imparted color.

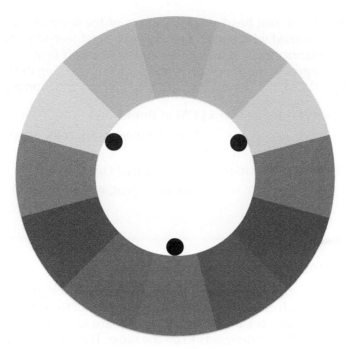

FIGURE 7.15
Triadic color scheme.

7.4 Color and Weave Effect

Woven fabrics can be produced with different color effects by using a combination of a particular weave and two or more colors with their special arrangement. *Callaway Textile Dictionary* defines *color and weave effect* as

> "A woven pattern in two or more colors produced by a combination of color and weave, but in appearance usually quite distinct from either the order of coloring or the weave."

From this definition, one can conclude that the effect in the cloth is dependent on three parameters, those are,

1. Weave design
2. Warp color arrangement
3. Weft color arrangement.

The above three parameters must be known to get a desired color pattern. A color pattern produced by this way is different in appearance from the arrangement of colors in warp and weft and also from the weave itself. This

is due to the weave that breaks the continuity of the warp and weft colors (float length is limited with respect to each weave) and secondly the color on the face of the fabric can be shown either by warp or weft float.

The weave repeat of the effect can be determined by taking the least common multiple (LCM) of the weave repeat and color combination repeat.

Ends of color pattern = LCM of threads in

(weave repeat, warp color plan)

Picks of color pattern = LCM of threads in

(weave repeat, weft color plan)

7.5 Representation on a Graph Paper

Color and weave effects are conveniently produced on the graph paper by the designers, which enable them to have the insight of the effect of various weaves with the same color arrangement. There are several CAD programs that are also used by the designers to do the same. These CAD programs are discussed in detail in Chapter 2. Figure 7.16 illustrates the representation of color and weave effects on a graph paper, with the indication of the weave design, the warp and weft color arrangement, and the resultant color pattern produced by these. By convention, colors are represented as dark D, medium M, and light L on the graph paper.

The steps to represent a color and weave effect on a graph paper are as follows:

1. First of all, indicate and fill the base weave.
2. Represent the warp colors and fill up the warp color where a warp yarn is above a weft yarn.

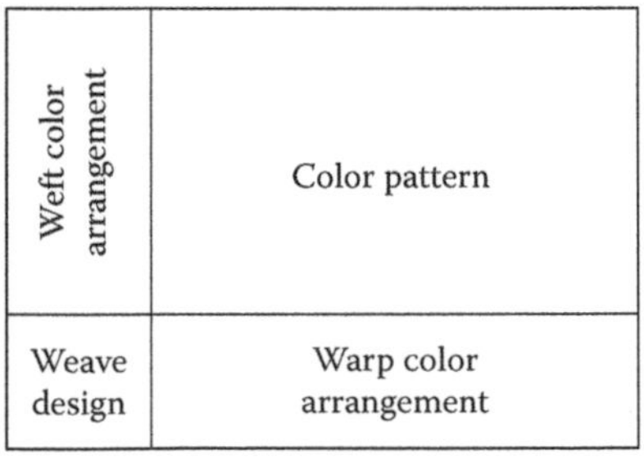

Weft color arrangement	Color pattern
Weave design	Warp color arrangement

FIGURE 7.16
Color and weave effect representation.

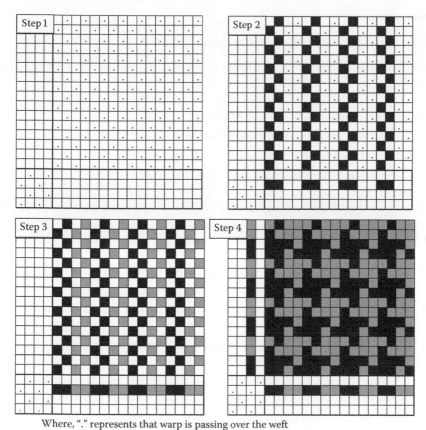

Where, "." represents that warp is passing over the weft

FIGURE 7.17
Steps to get a color and weave effect.

3. Represent the weft colors and fill up the weft color where a weft yarn is above a warp yarn.

For example, as shown in Figure 7.17, there is a bird's-eye effect with the base weave design of plain weave and the warp and weft color arrangements of 2D-2L with having effect repeat on 4 ends and 4 picks.

The same steps are followed for more than two color patterns. It is suitable to indicate more than one repeat of small color patterns so that the pattern becomes clearly visible. Different color patterns can be achieved by using the same weave and color arrangement either by starting the weave at different points with respect to warp and weft colors or by color arrangement instigating in a different manner with each other.

7.6 Classification of Color Patterns

The coloring arrangement in the warp and weft can be of the following types:

1. Simple warping and simple wefting
2. Simple warping and compound wefting
3. Compound warping and simple wefting
4. Compound warping and compound wefting

To each of the above arrangements, simple, stripe, and check weaves can be applied. A few terms related to the color and weave effect are defined in the following.

- *Simple warp*: Color combination in warp remains same, for example, 1L-2D.
- *Simple weft*: Color combination in weft remains same, for example, 1L-2D.
- *Compound warp*: Color combination in warp varies, for example, 1L-2D-1L-2D-2D-2L-2D-2L can also be represented as 2(2L-2D)-2(2D-2L).
- *Compound weft*: Color combination in weft varies, for example, 1L-2D-1L-2D-2D-2L-2D-2L can also be represented as 2(2L-2D)-2(2D-2L).
- *Simple weave*: Weave remains same, for example, 2/2 twill.
- *Stripe weave*: Weave is combination of two weaves, for example, 2/2 twill + 2/2 twill herringbone.
- *Check weave*: Weave is combination of four weaves, for example, 2/2 S twill + 2/2 Z twill + 3/1 S twill + 1/3 S twill.

7.7 Color Effects with Simple Pattern

Effects produced by simple warping, simple wefting, and simple weave are shown in Figure 7.18.

7.7.1 Solid Hairlines

Solid hairlines can be vertical or horizontal lines. Hairline is used to represent that each line is equal to the width of one yarn. Solid lines of greater width can also be produced. If the warp and weft color combination is same, the resultant effect will be horizontal lines, and if the warp or weft color

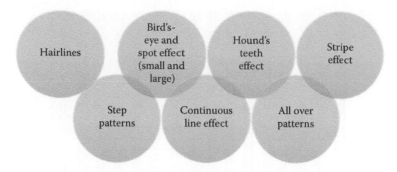

FIGURE 7.18
Color and weave effects with simple pattern.

combination is reciprocal to each other, vertical solid hairlines will be produced. Figures 7.19 and 7.20 show, respectively, horizontal and vertical solid hairlines with the plain weave. Same can be obtained with the 2/2 matt weave but with the strip width more than one yarn, as shown in Figures 7.21 and 7.22. With a 4-end sateen, horizontal hairlines are formed, as shown in Figure 7.23.

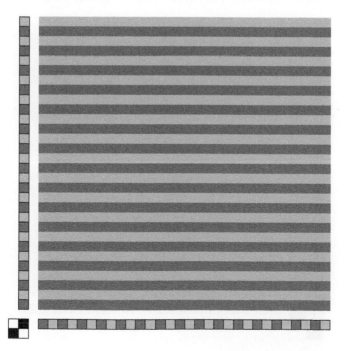

FIGURE 7.19
Horizontal solid hairlines, base weave: Plain weave, warp: 1D-1L, weft: 1D-1L.

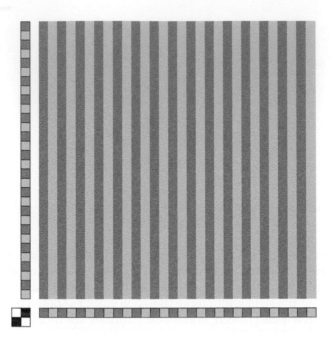

FIGURE 7.20
Vertical solid hairlines, base weave: Plain weave, warp: 1D-1L, weft: 1L-1D.

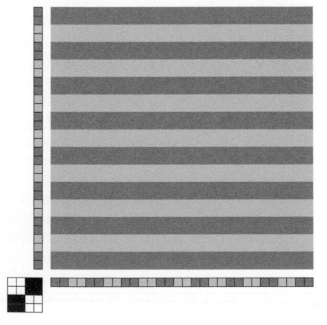

FIGURE 7.21
Horizontal hairlines, base weave: Basket weave, warp: 2D-2L, weft: 2D-2L.

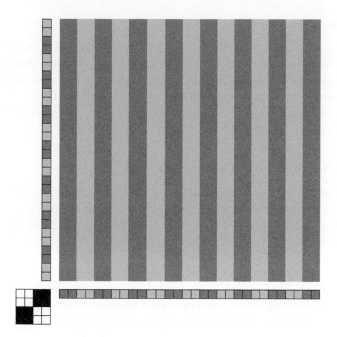

FIGURE 7.22
Vertical hairlines.

FIGURE 7.23
Horizontal hairlines, base weave: 4-end sateen, warp: 1D-1L, weft: 2D-2L.

7.7.2 Continuous Line Effect

Continuous line effect is a collection of three words, collectively meaning that this effect has continuous lines (horizontal or vertical) that have some effect as well. They are different from solid hairlines because of their effect. Examples of continuous line effect having the base weave of 2/2 twill with different combinations of warp and weft are shown in Figures 7.24 through 7.27 and with the 2/2 matt weave in Figures 7.28 and 7.29.

7.7.3 Shepherd's or Hound's Tooth Effect

This effect is somewhat similar to hound's teeth and that is where it gets its name from. Examples of hound's tooth effect are shown in Figures 7.30 and 7.31 with the 2/2 twill base weave, Figure 7.32 with the 3/1-1/3 fancy matt weave, and Figure 7.33 with the 2/2 herringbone.

7.7.4 Bird's-Eye or Spot Effect

In this effect, distinct, small, and separated from each other spots or patches of color appear on the whole surface of the fabric. Figures 7.34 through 7.36

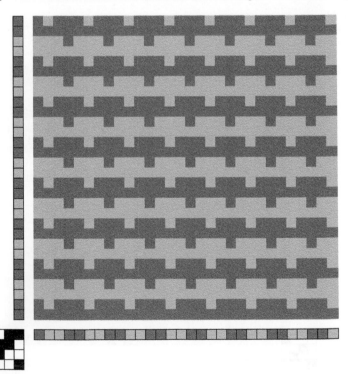

FIGURE 7.24
Horizontal continuous line effect, base weave: 2/2 twill, warp: 1D-(2L-2D), weft: 2D-2L.

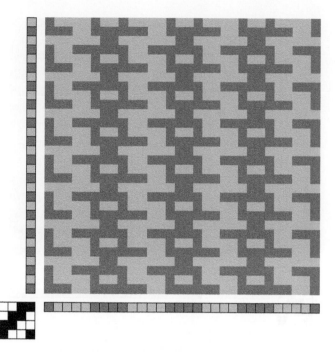

FIGURE 7.25
Vertical continuous line effect, base weave: 2/2 twill, warp: 1D-(4L-4D), weft: 1D-1L.

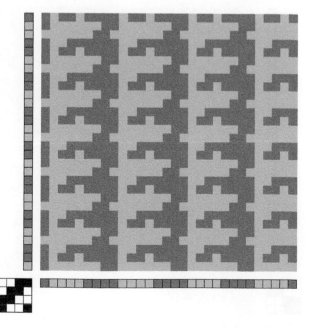

FIGURE 7.26
Vertical continuous line effect, base weave: 2/2 twill, warp: 1D-(4L-4D), weft: 1L-(2D-2L).

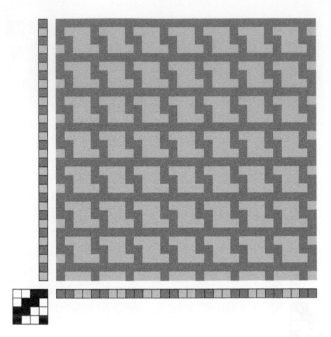

FIGURE 7.27
Horizontal and vertical continuous line effect, base weave: 2/2 twill, warp: 2D-2L, weft: 1L-1D.

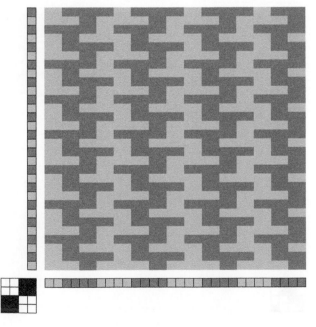

FIGURE 7.28
Vertical continuous line effect, base weave: 2/2 basket weave, warp: 2L-(4D-4L), weft: 1L-1D.

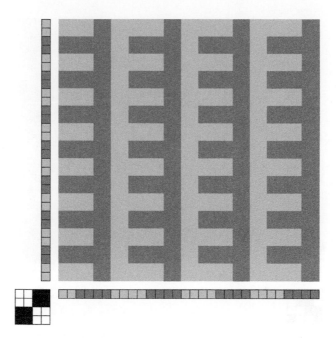

FIGURE 7.29
Vertical continuous line effect, base weave: 2/2 basket weave, warp: 2L-(4D-4L), weft: 2L-2D.

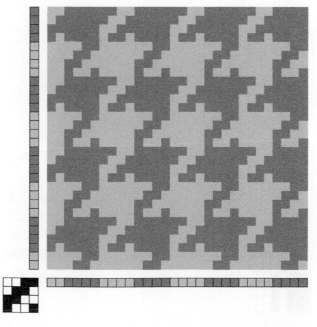

FIGURE 7.30
Hounds tooth effect, base weave: 2/2 twill, warp: 2L-(4D-4L), weft: 2L-(4D-4L).

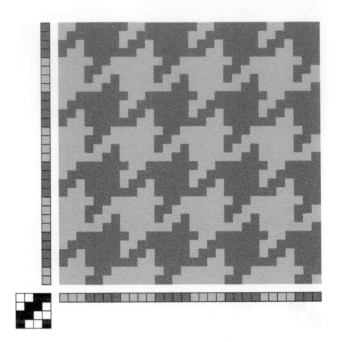

FIGURE 7.31
Hounds tooth effect, base weave: 2/2 twill, warp: 3L-(4D-4L), weft: 2L-(4D-4L).

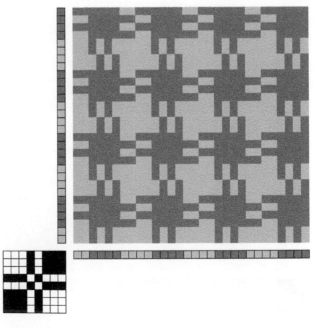

FIGURE 7.32
Hounds tooth effect, base weave: 3/1-1/3 fancy matt, warp: 2L-(4D-4L), weft: 2L-(4D-4L).

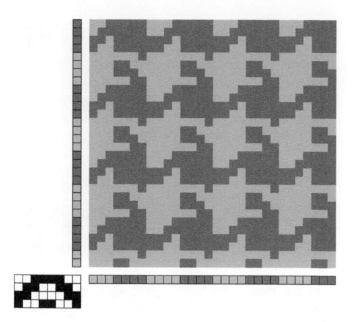

FIGURE 7.33
Hounds tooth effect, base weave: 2/2 herringbone, warp: 3L-(4D-4L), weft: 2L-(4D-4L).

show bird's-eye effect with the plain weave, Figures 7.37 and 7.38 with the 2/2 matt weave, Figures 7.39 and 7.40 with the 2/2 twill weave. Bird effect can also be formed with derived weaves. Figure 7.41 shows the effect with the 3/1-1/3 fancy matt weave and Figure 7.42 with the 3/1 herringbone.

7.7.5 Step Effect

In this effect, zigzag lines run diagonally on the surface of the fabric. This effect is conveniently produced with the twill weaves by arranging the color plan on half the threads of a weave repeat. Step effects with 2/1 twill and 2/2 twill are shown, respectively, in Figures 7.43 and 7.44, with 3/3 twill in Figures 7.45 and 7.46, and with 4/4 twill in Figures 7.47 and 7.48. Step effect can also be formed with the 2/2 matt weave, as shown in Figures 7.49 and 7.50.

7.7.6 All Over Effect

These color effects run all over the surface of the fabric. All over effects are best produced when the color plan and weave repeat are arranged on multiples of each other and require two or more repeats of each to produce one repeat of the effect. All over effect with 4/4 twill is shown in Figure 7.51, with 3/3 twill in Figure 7.52, and with 2/2 twill with different combinations of warp and weft in Figures 7.53 and 7.54.

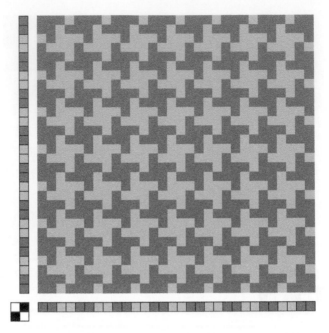

FIGURE 7.34
Bird's eye effect, base weave: Plain weave, warp: 2D-2L, weft: 2D-2L.

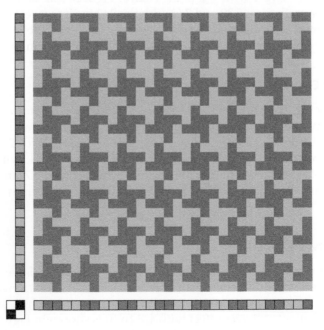

FIGURE 7.35
Bird's eye effect, base weave: Plain weave, warp: 1L-(2D-2L), weft: 1L-(2D-2L).

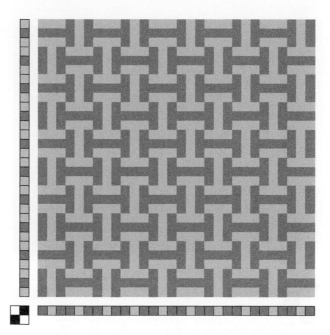

FIGURE 7.36
Bird's eye effect, base weave: Plain weave, warp: 1D-1L-1D, weft: 1L-1D-1L.

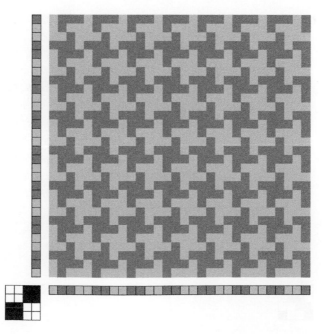

FIGURE 7.37
Bird's eye effect, base weave: 2/2 basket weave, warp: 1L-(2D-2L), weft: 1L-(2D-2L).

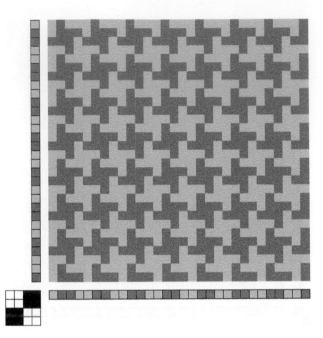

FIGURE 7.38
Bird's eye effect, base weave: 2/2 basket weave, warp: 1L-(2D-2L), weft: 1D-(2L-2D).

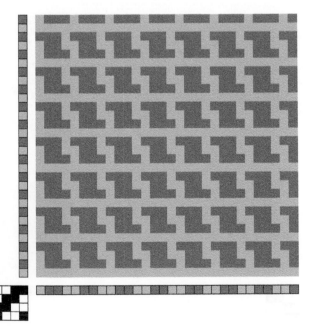

FIGURE 7.39
Bird's eye effect, base weave: 2/2 twill, warp: 1L-(2D-2L), weft: 1L-1D.

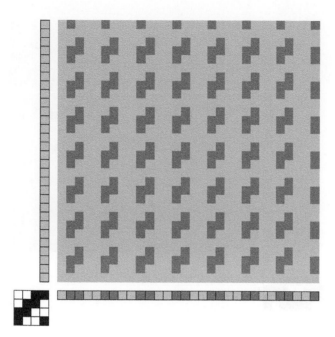

FIGURE 7.40
Bird's eye effect, base weave: 2/2 twill, warp: 1L-(2D-2L), weft: 1L.

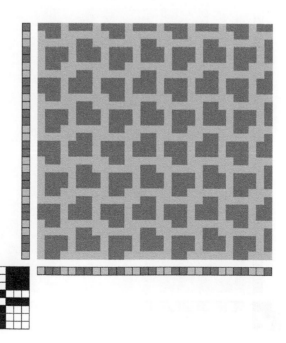

FIGURE 7.41
Bird's eye effect, base weave: 3/1-1/3 fancy matt, warp: 1L-(2D-2L), weft: 1L-(2D-2L).

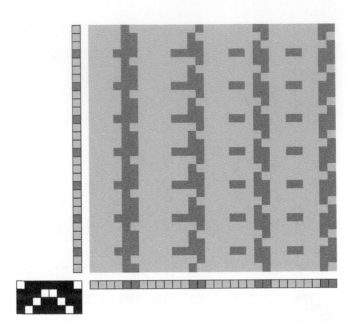

FIGURE 7.42
Bird's eye effect, base weave: 3/1 herringbone, warp: 4L-(2D-6L), weft: 2L-(1D-3L).

FIGURE 7.43
Step effect, base weave: 2/1 twill, warp: 1D-1L, weft: 1D-1L.

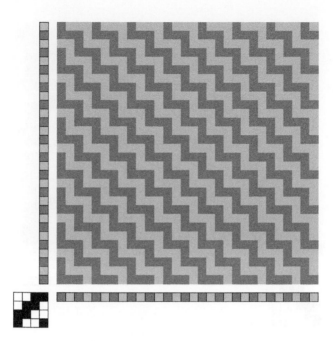

FIGURE 7.44
Step effect, base weave: 2/2 twill, warp: 1D-1L, weft: 1D-1L.

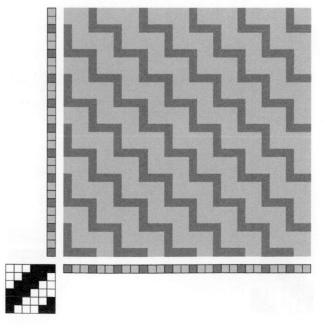

FIGURE 7.45
Step effect, base weave: 3/3 twill, warp: 1D-2L, weft: 1D-2L.

FIGURE 7.46
Step effect, base weave: 3/3 twill, warp: 1D-1M-1L, weft: 1D-1M-1L.

FIGURE 7.47
Step effect, base weave: 4/4 twill, warp: 2D-2L, weft: 2D-2L.

FIGURE 7.48
Step effect, base weave: 4/4 twill, warp: 1M-1D-2L, weft: 1M-1D-2L.

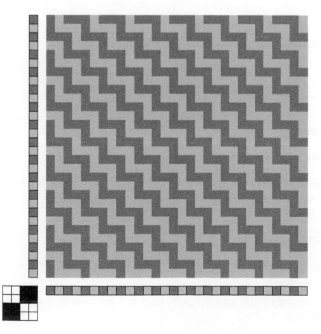

FIGURE 7.49
Step effect, base weave: 2/2 basket weave, warp: 1D-1L, weft: 1L-1D.

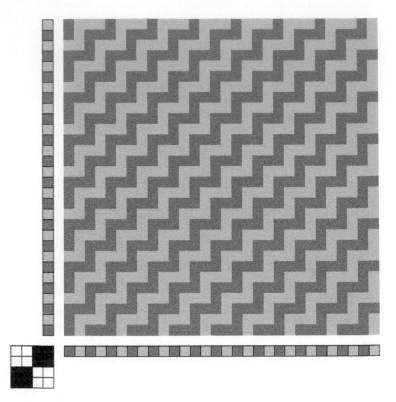

FIGURE 7.50
Step effect, base weave: 2/2 basket weave, warp: 1D-1L, weft: 1D-1L.

7.8 Stripes and Checks

Stripes and check effects can be formed with combinations of simple and compound warping and wefting, as shown in Figure 7.55.

7.8.1 Joining of Weaves

- Avoid formation of long floats, where different sections of designs are in contact.
- Equal-sided twills and weaves that are reverse of each other can be joined.
- Plain weave or weave of minimum float length can be introduced between two weaves.

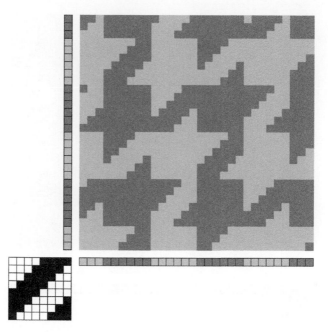

FIGURE 7.51
All over effect, base weave: 4/4 twill, warp: 3L-(6D-6L), weft: 3L-(6D-6L).

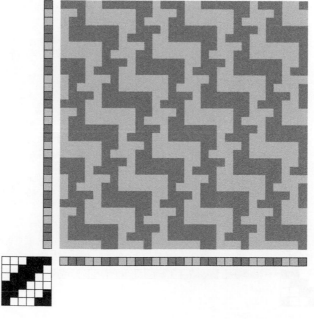

FIGURE 7.52
All over effect, base weave: 3/3 twill, warp: 1L-(2D-2L), weft: 1L-(2D-2L).

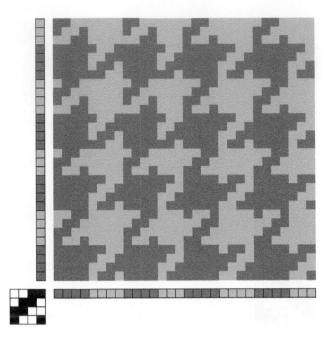

FIGURE 7.53
All over effect, base weave: 2/2 twill, warp: 4D-4L, weft: 4D-4L.

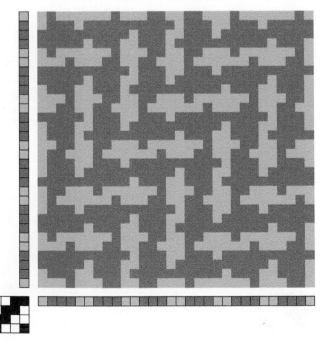

FIGURE 7.54
All over effect, base weave: 2/2 twill, warp: 1L-(3D-2L), weft: 1L-(3D-2L).

7.8.2 Combination of Weaves

- Same weave usually a twill but turned in opposite direction
  - For example, 3/1 S twill and 3/1 Z twill
- Sections of different weaves derived from the same base weave
  - For example, 4 sections of 2/2 pointed, 2/2 herringbone, 2/2 skip, and 2/2 twill
- Combination of warp and weft face weaves
  - For example, 4-end satin and 4-end sateen
- Combination of different weaves
  - For example, plain weave, 2/1 twill, 4-end satin, 2/2 matt

Figures 7.56 through 7.67 show different stripes and check effects with different combinations as shown in Figure 7.55.

Order of coloring	Simple weave	Stripe weave	Check weave
Simple warping and simple wefting	Simple pattern	Stripe pattern	Check pattern
Compound warping and simple wefting	Stripe pattern	Stripe pattern	Check pattern
Simple warping and compound wefting	Crossover pattern	Check pattern	Check pattern
Compound warping and compound wefting	Check pattern	Check pattern	Check pattern

FIGURE 7.55
Simple, stripe, and check pattern combinations.

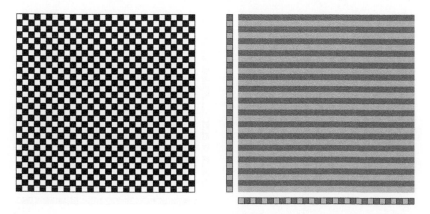

FIGURE 7.56
Simple pattern, base weave: Plain weave, warp: 1L-1D, weft: 1L-1D.

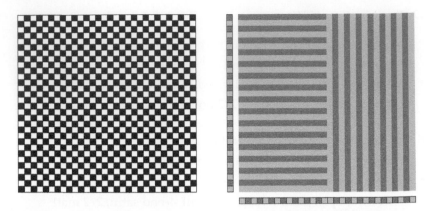

FIGURE 7.57
Stripe pattern, base weave: Plain weave, warp: 7(1L-1D)-2L-7(1D-1L), weft: 1L-1D.

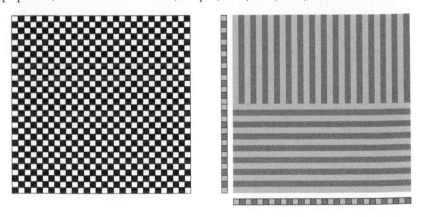

FIGURE 7.58
Stripe pattern, base weave: Plain weave, warp: 1L-1D, weft: 7(1L-1D)-2L-7(1D-1L).

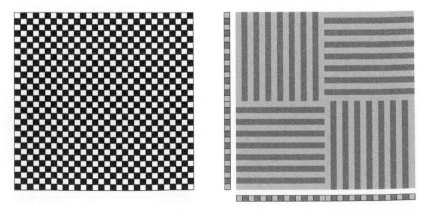

FIGURE 7.59
Check pattern, base weave: Plain weave, warp: 7(1L-1D)-2L-7(1D-1L), weft: 7(1L-1D)-2L-7(1D-1L).

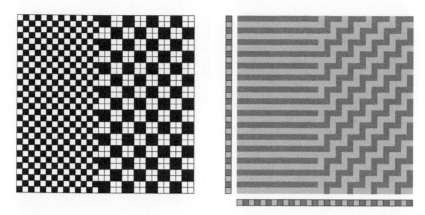

FIGURE 7.60
Stripe pattern, base weave: Plain weave and its derivatives, warp: 1L-1D, weft: 1L-1D.

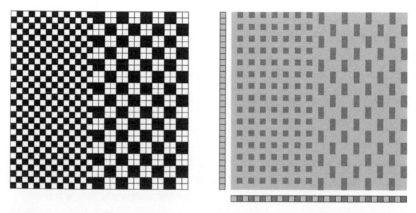

FIGURE 7.61
Stripe pattern, base weave: Plain weave and its derivatives, warp: 1L-1D, weft: 1L.

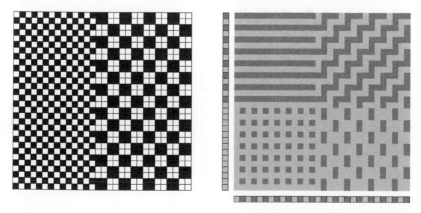

FIGURE 7.62
Check pattern, base weave: Plain weave and its derivatives, warp: 1L-1D, weft: 14(1L)-8(1L-1D).

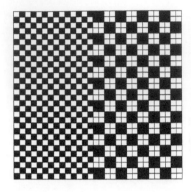
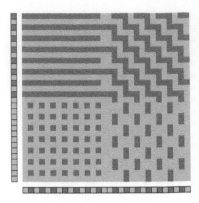

FIGURE 7.63
Check pattern, base weave: Plain weave and its derivatives, warp: 7(1L-1D)-2L-7(1D-1L), weft: 14(1L)-8(1L-1D).

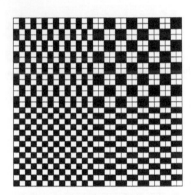
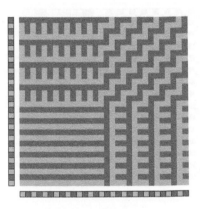

FIGURE 7.64
Check pattern, base weave: Plain weave and its derivatives, warp: 1L-1D, weft: 1L-1D.

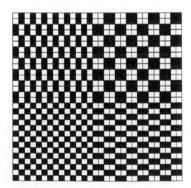
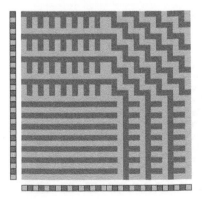

FIGURE 7.65
Check pattern, base weave: Plain weave and its derivatives, warp: 7(1L-1D)-2L-7(1D-1L), weft: 1L-1D.

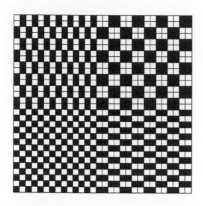 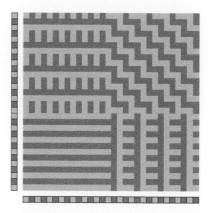

FIGURE 7.66
Check pattern, base weave: Plain weave and its derivatives, warp: 1L-1D, weft: 6(1L-1D)-2L-8(1D-1L).

 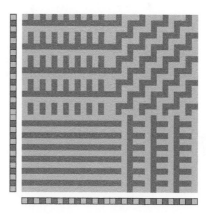

FIGURE 7.67
Check pattern, base weave: Plain weave and its derivatives, warp: 7(1L-1D)-2L-7(1D-1L), weft: 6(1L-1D)-2L-8(1D-1L).

References

Bohren, C. F. and E. E. Clothiaux. Absorption: The death of photons: Fundamentals of atmospheric radiation. *An Introduction with 400 Problems*, Wiley, Hoboken, New Jersey, United States, 2006.

Judd, D. B. and G. Wyszecki. *Color in Business, Science and Industry*, Wiley, Hoboken, New Jersey, United States, 1963.

Swenson, R. Optics, Gender, and the Eighteenth-Century Gaze: Looking at Eliza Haywood's Anti-Pamela. *The Eighteenth Century*, 51(1), 2010, 27–43. Project MUSE, doi:10.1353/ecy.2010.0006.

Section II

Knitted Fabric Structures

8

Introduction to Knitting

Hafiz Shehbaz Ahmad

CONTENTS

8.1 History of Fabric

Clothing is one of the three basic needs of human beings. At the very start of human presence on the planet Earth, humans started to explore a better method and material to protect their body from extreme weather conditions. They used barks, leaves, and hides of animals to protect their body. Felting is reported to be the first fabric-forming technique dating back to roughly 100,000 years ago.

Textile fabric is defined as a 2D plane-like structure made of textile fibers, yarns, or filaments by weaving, knitting, braiding, nonwoven, or any other technique.

8.2 History of Knitting

The concept of knitting from converting yarn into fabric using loops dates back to some 3000 years ago. Weft knitting is the first ever used technique to knit a fabric by means of two handheld pins. Hand pin knitting dates back to 1350 in religious paintings of Northern Italy; it then spread through the rest of Europe. Cap knitting was established in Britain by 1424, men first started knitting as a profession in the fifteenth century, and the parliament controlled the prices of knitting caps in 1488. In 1589, Reverend William Lee invented the hand-frame knitting, which changed the conventional method of hand knitting to semi-mechanization. William started his efforts in 1561 and worked for the needles to work on a faster rate. The frame was perfected during the period of time, and it was commercialized in 1589.

Knitting is the second largest and oldest technology of fabric manufacturing after weaving [1]. In this technique, a continuous length of yarn is converted into a fabric by interlooping. The process of interlooping is done either by hand or by a machine. Minimum one yarn or one set of yarn is required to get converted into a fabric. Knitted fabrics have advantage over woven fabrics in terms of cost, process ability, flexibility of design, and other performance properties. Depending upon the direction of the yarn feeding and and the direction of fabric formation, knitting can be categorized into two types: warp knitting and weft knitting. If the direction of yarn feeding and the direction of fabric formation are parallel to each other, then the type of knitting is called the warp knitting, and if the direction of yarn feeding and the direction of fabric formation are perpendicular to each other, then the process of knitting is called the weft knitting. Out of these two types, the weft knitting is widely used in Pakistan and India [1].

8.3 Weft Knitting

If the direction of yarn feeding is perpendicular to the direction of fabric formation or if yarn is supplied in a weft or crosswise direction, then this type of knitting is called the weft knitting (Figure 8.1). The fabrics formed through this type of mechanism are called weft-knitted fabrics or simple jersey fabrics and the machine used is called the weft knitting machine.

Weft knitting machines may be flat or in a circular form. Mostly latch needle are used in weft knitting. Both natural and synthetic yarn can be used in weft knitting. There are grooves or cuts on the cylinder or dial of machines at regular intervals. The number of cuts in unit length is called the gauge of

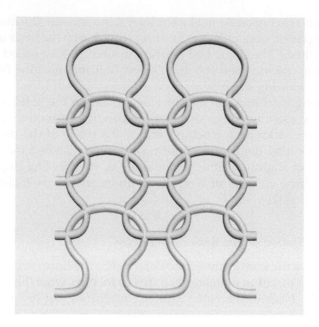

FIGURE 8.1
Basic structure of weft knitting.

the machine. The finer the gauge, the more will be the number of needles per unit area. On the basis of needle beds, the machines are classified as single bed or double bed.

8.3.1 Classification of Circular Weft Knitting

Weft knitting machines can be classified in the following ways:

- On the basis of the number of needle beds
- On the basis of diameter
- On the basis of the nature of driving system
- On the basis of knitted structure
- On the basis of design elements
- On the basis of special products

8.3.1.1 Classification on the Basis of the Number of Needle Beds

The basic element for the knitting machine classification is the needle bed. The needle bed is the main part of a knitting machine and provides the platform for the movement of needles. It has cuts or grooves that provide the direction of movement to the knitting needles. This direction may be up and

down or to and fro but could not be the lateral movement. The machine may be called single-bed or double-bed knitting machine on the basis of the number of needle beds. The single-bed knitting machine is also called the single-jersey knitting machine and the double-bed knitting machine is called the double-jersey knitting machine.

Weft knitting machines can also be categorized on the basis of the shape of the knitting machine. If the shape of the needle bed is circular, then it is called the circular knitting machine, and if the shape of the needle bed is flat, then it is called the flat-bed knitting machine. Flat-bed machines may be horizontal or slightly inclined at an angle of 90–105°. The single circular knitting machine consists of a single cylinder, which is inclined in the vertical direction [2].

8.3.1.2 Classification on the Basis of Diameter

The diameter of the knitting machine determines the linear width of knitted fabrics. So, the machine diameter has direct relation with the fabric linear width. On the basis of diameter, knitting machines can be classified into three categories:

- Small-diameter knitting machine (diameter ranges from 3 to 6 inches)
- Medium-diameter knitting machine (diameter ranges from 8 to 22 inches)
- Large-diameter knitting machine (diameter ranges from 24 to 40 inches)

Small-diameter machines are used for hosiery products, medium-diameter machines are used for body size fabrics, and large-diameter machines are used for the production of open width fabric similar to the one produced on flat knitting machines.

8.3.1.3 Classification on the Basis of the Nature of Driving Systems

The driving mechanism is also a base for the classification of knitting machines. On the basis of the nature of driving mechanism, knitting machines can be divided into two categories:

1. Hand-driven knitting machine
2. Power-driven knitting machine

Both these types of knitting machines are considered under the list of flat knitting machines.

8.3.1.4 Classification on the Basis of Knitted Structures

Knitting machines can be categorized on the basis of knitted structures. There are four basic knitted structures, that is, single jersey, rib, interlock, and purl. The single-jersey fabric is made on single knit machines while other three structures are developed on double knit machines.

8.3.1.5 Classification on the Basis of Design Elements

Sometimes, extra design elements are also attached on the knitting machines for the sake of designing, and these machines are named after design elements. For example,

- Pattern wheel knitting machine
- Jacquard knitting machine
- Multitrack knitting machine
- Intarsia knitting machine
- Knitting machine with CAD and CAM

8.3.1.6 Classification on the Basis of Special Products

An end product is also a base for the classification of knitting machines. Knitting machines are named on the basis of products obtained from that machine. Examples include socks knitting machine, terry knitting machine, gloves knitting machine, sliver knitting machine, and fleece knitting.

8.4 Warp Knitting

If the direction of yarn feeding is parallel to the direction of fabric formation or if yarn is supplied in a warp-wise direction, then this type of knitting is called the warp knitting (Figure 8.2). The process of yarn feeding of warp knitting is just like feeding of warp yarn in weaving. The fabrics formed through this type of mechanism are called warp-knitted fabrics and the machine used is called the warp knitting machine. There is a difference between the loop formation process of warp knitting as compared to weft knitting. In weft knitting, needles move sequentially and a single yarn is converted to a sequence of loops to form a complete course. But in warp knitting, a set of needles are activated to form a series of loops at the same time. Warp knitting machines are further classified into tricot and Raschel machines on the basis of their structural mechanisms.

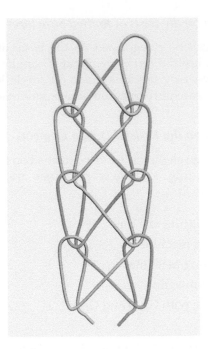

FIGURE 8.2
Basic structure of warp knitting.

Initially, both tricot and Raschel knitting were categorized on the basis of the type of needles used in each machine; tricot machines used bearded needles and Raschel machines used latch needles. But later on, compound needles replaced the bearded needles in tricot knitting machines. So, the categorization of warp knitting machines on the basis of needles is no longer possible. Now, the warp knitting machines are categorized on the basis of the type of sinkers used. The sinker function is different for both types. In tricot machines, the function of sinkers is to hold the fabric throughout the knitting cycle, whereas in Raschel machines, the function of sinkers is to ensure the stay of fabric while the needle moves upward.

8.5 Knitting Elements of Weft Knitting

There are three major knitting elements in weft knitting.

8.5.1 Needle

There are three types of needles used in weft knitting, which are, latch, bearded, and compound needles. Latch needle is mostly used in weft and

warp knitting. Irrespective of the size of the needle, each latch needle has the following major parts (Figure 8.3):

Hook: It has the major role in yarn catching and holding it. It is in the curved form and the yarn is trapped in it. The top portion of a hook is known as crown.

Latch: The name of the latch needle is due to this part. The major function of a latch is to hold the yarn within a hook.

Rivet: Rivet plays the role of fulcrum and fixes the latch with the stem of the needle.

Butt: Butt is the bouffant part of the needle. The needle moves in cam with the help of a butt.

Tail: Tail is the lower most part of the needle below the butt. It provides support to the needle during its movement in cam.

Major advantage of using latch needles is that they are robust and self-sufficient, and they need no extra element for hook closing and opening. The latch needle is not suitable for finer gauge machine.

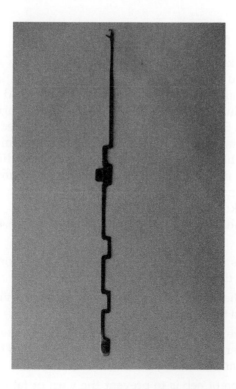

FIGURE 8.3
Latch needle.

8.5.1.1 Knitting Actions of Latch Needle

1. *Idle position*: In idle position, the yarn is in the hook of the needle and the needle is at the top of the verge of the sinker.
2. *Latch opening*: In latch opening process, the needle moves upward direction and the yarn from the previous loop moves downward and opens the latch of the needle.
3. *Clearing height*: When the needle reaches its maximum height, the old loop is removed from the latch and the latch moves downward.
4. *Yarn feeding*: The needle starts moving downward direction and the hook of the needle catches the new yarn. During the downward movement of the needle, the previous loop closes the latch and rides over the latch. At the same time, the sinker starts moving backward direction.
5. *Casting off*: The process of sliding off the old loop from the needle is called casting off. As the needle descends, the previous cleared loop moves outward direction of the needle and the loop slides off, which is called the cast-off loop. The sinker continues its outward movement and the old loop lies on the throat of the sinker.

8.5.2 Sinker

A sinker is a basic knitting element next to a needle. It is a thin metal plate, which is positioned between needles. Sinkers are used in both warp and weft knitting machines. In weft knitting, sinkers are used in single cylinder machines. The main function of sinkers is to hold the fabric during the loop formation process. Sinkers are placed in horizontal direction and move to and fro in horizontal direction. Both needle and sinker move at right angle to each other. The shape and function of a sinker may be different depending upon the machine manufacturer. Irrespective of the manufacturer, sinkers have to perform one of the following functions: (a) holding loop, (b) loop formation, and (c) knocking over.

A sinker of any shape has the following parts (Figure 8.4):

Throat: Throat is the main part of a sinker. Its function is to hold the fabric during the loop formation.

Belly: Belly is the projected portion of a sinker. Previous loops and fabrics rest on the belly.

Butt: Just like the needle, butt is the part of a sinker, which is responsible for sinker movement. Butt receives drive from cam and moves the sinker to and fro along the horizontal direction.

Neb: The function of neb is to prevent the yarn or fabric from moving along the needle during its upward motion.

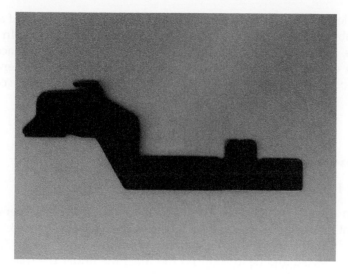

FIGURE 8.4
Sinker for single knit machine..

8.5.3 Cam

Knitting cam is the third most important element of knitting machines. Cam is a metal plate, which has channels or grooves, and provides path to a needle during the loop formation process. The butt of a needle moves in cam and the knitting action takes place due to the shape of a cam (Figure 8.5).

Cams are fixed in a cam box and each cam is fitted in a specific cam track. Needles are placed in a specific track with respect to their butt position.

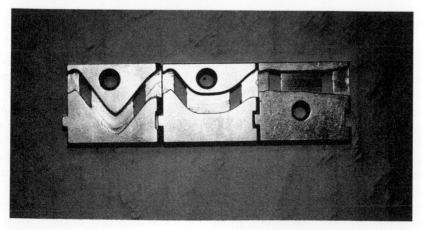

FIGURE 8.5
Knit, Tuck, and Miss cams.

The profile of a cam defines the height of a needle and the shape of a loop in knitted fabrics [3]. The height of a cam also defines the loop length. In modern machines, previously used linear cam systems have been replaced with nonlinear cam systems. In the linear cam system, there is a sudden change in the velocity of a needle, which has speed limitations and chance of needle breakage.

8.6 Flat Knitting Machine

American clergyman Rev. Isaac William Lamb was the first person who got patent of the hand-powered flat knitting machine in 1863. In 1864, some modification in Lamb's flat knitting machine was done by American Henry J. Griswold and he introduced the machine under the trade name of "Climax," "Crane," and "Little Rapid." These preliminary machines were used in cottage industries for the production of socks and children wears. These machines laid the foundation for the development of circular knitting machines. Henri Eduard Dubied took notice of the Lamb machine in the world Expo in Paris in 1867 and got the patent for the production of knitting machines in his Swiss machinery unit.

In 1867, Dubied introduced his first flat-bed knitting machine named as "Trikoteuse Omnibus." Narrow diameter machines were developed by Americans as well as Europeans for the production of socks. The knitting industry began to expand in Germany and France at the end of 1860. Laue und Timaeus, a German company, is the first commercial flat knitting machine manufacturer on the basis of the Lamb system.

8.6.1 Types of Flat-Bed Knitting Machines

Flat knitting machines are the second type of weft knitting. The yarn feeding and working principle are the same as those of circular knitting machines. Flat knitting machines can be categorized on the basis of number of beds and shape of needle beds.

On the basis of the shape of a needle bed, flat knitting machines are divided into V-bed and links-links flat knitting machines. The V-bed knitting machines consist of two beds that are vertically aligned with each at the angle of 90–105° in such a way that gives the shape of V. The V-bed machines are rib gated. Interlock V-bed machines are not feasible because needles of two beds are difficult to manage and there will be a chance of their collaboration. The needle bed in the direction of a viewer is called front needle bed and the needle bed in the opposite direction to the viewer is called back needle bed or rear needle bed. Both needle beds can be used simultaneously and in this way a double faced fabric just like the double knit

circular knitted fabric is formed. By using only one needle bed, a single knit fabric just like single jersey is made on V-bed machines.

In links-links flat knitting machines, both the needle beds are horizontally in front of each other and both the beds are flat instead of inclination at an angle. Another difference is the use of a purl or floating needle, which is able to produce stitches on either side.

The purl needle is controlled by two jacks, either of which is in use during the knitting, enabling the work to be transferred automatically from a purl to a plain or a plain to a purl in any combination selected. When the needle is coupled to the jack, the stitches may be automatically transferred to the opposite bed, a purl to a plain or a plain to a purl, or double transferred to reverse the order of the stitches from the previous row knitted.

References

1. S. R. Chan, *Fundamentals and Advances in Knitting Technology*. New Delhi: Woodhead Publishing India Pvt Ltd, 2011.
2. C. Mazza, *Knitting*, Milano, Italy: dell'Associazione Costruttori Italiani di Macchinario per l'Industria Tessile (ACIMIT), 2001.
3. J. J. F. Knapton and D. L. Munden, A study of the mechanism of loop formation on weft-knitting machinery: Part I: The effect of input tension and cam setting on loop formation. *Text. Res. J.* 36(12), 1072–1080, 1966.

9

Patterning in Weft Knitting

Waqas Ashraf

CONTENTS

9.1 Principle Stitches in Weft Knitting

There are four basic types of stitches in weft knitting, namely knit, tuck, purl, and miss or float. Mostly, weft knitted fabrics and their derivatives are based on the combination of these stitches.

9.1.1 Knit Stitch

This knit stitch is formed when the needle is raised enough to engage new yarn in the hook by the camming action and old loop is cleared. The technical back side of the knit stitch is called a purl stitch. The knit appearance of face and back sides are shown in Figure 9.1. The clearing position of knit stitch is shown in Figure 9.2a.

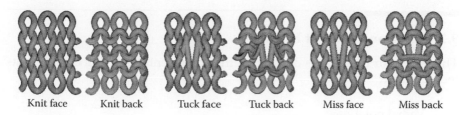

Knit face   Knit back   Tuck face   Tuck back   Miss face   Miss back

FIGURE 9.1
Basic weft knitting stitches pattern.

9.1.2 Tuck Stitch

Tuck stitch is formed when needle is raised to get new yarn but not enough to clear the previous formed loop. The needle then holds two loops when it descends as shown in Figure 9.2b. The needle can hold up to four loops, so four consecutive tuck stitches can be formed in a wale. The fabric gets thicker with tuck stitch as compared to knit stitches due to the accumulation of yarns. The needle clears all the old loops at knit stitch. The structure becomes more open and permeable to air than knit stitches. It can also be used to get different color effects in fabric. The tuck stitch patterns from face and back sides are shown in Figure 9.1.

9.1.3 Miss or Float Stitch

When the needle does not move upward to clear the old loop and also does not take the new yarn that is presented to it, then miss or float stitch is formed. Needle is not activated in miss stitch. Moreover, it holds the old loop as shown in Figure 9.2c. Float stitch on the successive needles produces longer float of yarn, which may cause the problem of snagging. The float is preferably used where we need to hide color yarn from the technical face of the fabric. The hidden yarn floats at the back side of the fabric as shown in Figure 9.1. The yarn gets straighter in float stitch construction, so the extensibility decreases as compared to tuck and knit stitches.

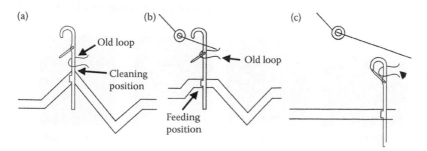

FIGURE 9.2
Latch needle position for basic stitches.

9.2 Notations in Weft Knitting

There are four basic types of notation that represent the design of weft knitted structure:

1. Verbal notation
2. Graphical notation
3. Symbolic notation
4. Diagrammatic notation

9.2.1 Verbal Notation

In this technique of notation, the design is represented verbally, for example, the design has two courses repeat and has a combination of knit and tuck stitches at alternate positions. This notation technique does not give enough information about the design. This technique is not feasible for complex structures.

9.2.2 Graphical Notation

In this technique, the design is represented graphically by line diagram. This technique gives the best idea of the design structures and stitch pattern. Representation of design using this technique is a time-consuming process, and skill may require if the design is constructed by means of any software. The graphical representations of plain jersey and Rib 1×1 are given in Figure 9.3.

9.2.3 Symbolic Notation

In this technique, the individual stitch is repeated by columns and rows pattern. The rows represent the individual course, and the columns represent the wales. To represent the type of stitches, following symbols are used:

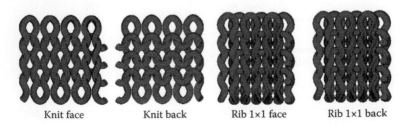

Knit face   Knit back   Rib 1×1 face   Rib 1×1 back

FIGURE 9.3
Graphical representation of plain jersey and rib structure.

	W1	W2	W3	W4
C1	X	X	X	X
C2	X	X	X	X
C3	X	X	X	X
C4	X	X	X	X

Single jersey

	W1	W2	W3	W4
C1	X	X	X	X
C2	X	•	X	•
C3	X	X	X	X
C4	•	X	•	X

Lapique

	W1	W2	W3	W4
C1	X	O	X	O
C2	X	O	X	O
C3	X	O	X	O
C4	X	O	X	O

Rib 1×1

FIGURE 9.4
Symbolic representation of desing.

Knit = $\boxed{X}$

Tuck = $\boxed{\cdot}$

Miss = $\square$

Purl = $\boxed{O}$

The symbolic notations of single jersey, lapique and Rib 1×1 are shown in Figure 9.4. The W and C represent the wales and courses, respectively.

9.2.4 Diagrammatic Notation

This technique to represent the design is easy to understand, and it gives enough information about the design. This technique is also helpful to understand the double-knit structure. Each yarn in this technique represents the individual stitch both on dial and cylinder. The diagrammatic notations of different designs are shown in Figure 9.5.

The stitch types are represented in the following way:

A complete circle facing down represents the knit stitch.

A complete circle facing up represents the purl stitch.

A half circle or v-type notation shows the tuck stitch.

A float or straight line represents the miss of float stitch.

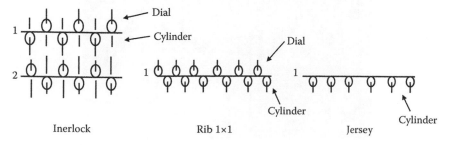

FIGURE 9.5
Diagrammatic representation of basic weft knit structure.

9.2.5 Cam Repeat Notation

The arrangement of cams in cam track system is very important to achieve the required design of the fabric. There are three basic types of stitches in weft knitting, so accordingly three different cams are placed in cam box. There are two things that need to be understood about cam notation. The first is feeders, and the second is tracks. Each feeder represents the individual course and tracks give the option to use different types of stitches within the same course. On single-knit machine, the cams are placed on cylinder side only, whereas the cams repeat for both dial and cylinder sides are considered to understand the design produced on interlock and rib machine. Figure 9.6 shows the actual image of cams of single-knit machine.

There are three notations used to represent the design:

Knit = ∧
Tuck = ⌐⌐
Miss = −

9.2.6 Needle Repeat Notation

The needle shows how the needles are placed inside tricks of the dial and cylinder sides. The needle repeat shows the needle number and tracks. Each needle represents the respective wale in the fabric. The needle has different butt positions for different tracks. From the top, the first track needle has the minimum hook-to-butt distance, and the needle for lower track has the

FIGURE 9.6
Cam track of single knit structure.

maximum hook-to-butt distance. If there are two or more tracks used on machine, the needles of different tracks are placed in dial and cylinder side according to the design repeat.

Needle active = |

Needle drop = ×

9.3 Single Jersey and Its Derivatives

9.3.1 Structure with One-Needle Type

There is a structure that can be produced with minimum one-needle type. Each course has only one type of stitches. Plain jersey fabric structure is one of the examples of single-needle-type design. This fabric can be produced on a minimum of one track. The design and cam repeat are shown in Figure 9.7.

Needle repeats of single-needle type of single-knit design are shown in Figure 9.8.

This design can be produced on maximum tracks available on machine in cam track lifting mechanism.

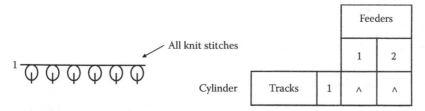

FIGURE 9.7
Design and cam repeat of single jersey.

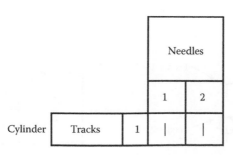

FIGURE 9.8
Needle repeat of single-needle-type single knit design.

		Needles	
		1	2
Cylinder	Tracks	1	$\vert$	
		2		$\vert$

FIGURE 9.9
Needle repeat of two-needle-type single knit design.

9.3.2 Structure with Two-Needle Type

These are the structures that require a minimum of two tracks to produce on machine. The needle arrangement is shown in Figure 9.9.

9.3.2.1 Cross Miss

This design has two courses repeat. This derivative is formed by the alternative position of knit and miss stitches. The cams repeat and cross miss design are shown in Figure 9.10.
  Design Pattern:

  Course 1: has knit stitches on odd needles only.

  Course 2: has knit stitches on even needles only.

9.3.2.2 Plain Pique

This design has two courses repeat. The design has a combination of tuck and knit stitches. The cam repeat and design of plain pique are shown in Figure 9.11.

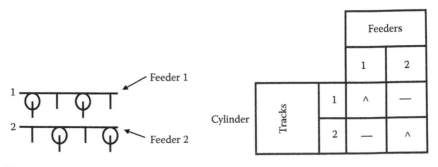

FIGURE 9.10
Cam repeat and design of cross miss.

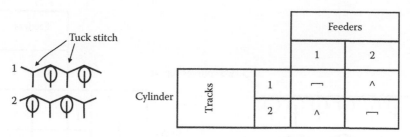

FIGURE 9.11
Cam repeat and design of plain pique.

Design Pattern:

Course 1: has knit stitches on even needles with tuck stitches on odd needles.

Course 2: has knit stitches on odd needles with tuck stitches on even needles.

9.3.2.3 Lapique

This design has four courses repeat formed by a combination of knit and tuck stitches. The cam arrangement and design of lapique are shown in Figure 9.12.
Design Pattern:

Course 1: has all the knit stitches.

Course 2: has knit stitches on even needles with tuck stitches on even needles.

Course 3: same as Course 1.

Course 4: has knit stitches on even needles with tuck stitches on odd needles.

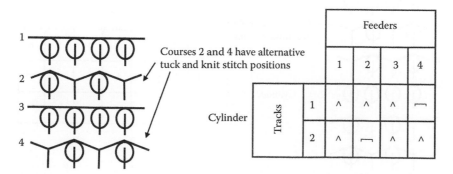

FIGURE 9.12
Cam repeat and design of lapique.

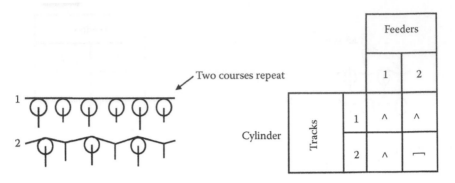

FIGURE 9.13
Cam repeat and design of longitudinal tuck stripe.

9.3.2.4 Longitudinal Tuck Stripe

This design has two courses repeat formed by a combination of knit and tuck stitches. The cam arrangement and design are shown in Figure 9.13.
Design Pattern:

Course 1: has all the knit stitches.

Course 2: has knit stitches on odd needles with tuck stitches on even needles.

9.3.2.5 Double 1 × 1 Cross Tuck

This design has six courses repeat formed by a combination of knit and tuck stitches. The cam arrangement and design are shown in Figure 9.14.

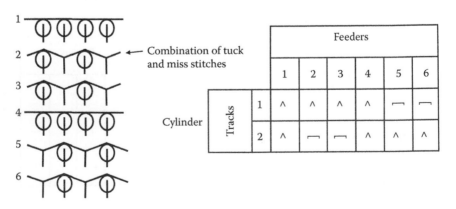

FIGURE 9.14
Cam repeat and design of longitudinal 1 × 1 cross tuck.

Design Pattern:

Course 1: has all the knit stitches.

Course 2: has knit stitches on odd needles with tuck stitches on even
needles.

Course 3: same as Course 2.

Course 4: same as Course 1.

Course 5: has knit stitches on even needles with tuck stitches on odd
needles.

Course 6: same as Course 5.

9.3.3 Fleece Structures

The fleece structure is formed by tucking the loop yarn with ground yarn
at selective position. This structure can be modified either by adopting the
different tucking positions of fleece yarn or by increasing the fleece yarn
courses in the structure repeat. Different fleece structure derivatives are
given in the following.

9.3.3.1 1 × 1 Fleece

The design has four courses repeat. The first and third courses represent the
ground yarn, while the second and fourth courses represent the fleece yarn.
The design has all types of stitches involved in this design. The cam arrangement and design are shown in Figure 9.15.

Design Pattern:

Course 1: has all the knit stitches.

Course 2: has all the tuck stitches on odd needles only.

Course 3: has all the knit stitches.

Course 4: has all the tuck stitches on even needles only.

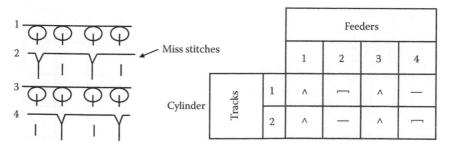

FIGURE 9.15
1 × 1 Fleece.

9.3.3.2 2 × 2 Fleece

The design has four courses repeat. The first and third courses represent the ground yarn, while the second and fourth courses represent the fleece yarn. The design has all types of stitches involved in this design. The cam arrangement and design are shown in Figure 9.16.

Design Repeat:

Course 1: has all the knit stitches.

Course 2: has a combination of two tuck and two float stitches, started with tuck stitches.

Course 3: has all the knit stitches.

Course 4: has a combination of two tuck and two float stitches, started with miss or float stitch.

9.3.3.3 3 × 1 Fleece

The design has four courses repeat. The first and third courses represent the ground yarn, while the second and fourth courses represent the fleece yarn. The design has all types of stitches involved in this design. The cam arrangement and design are shown in Figure 9.17.

Design Repeat:

Course 1: has all the knit stitches.

Course 2: has a combination of one tuck and three float stitches, started with tuck stitches.

Course 3: same as Course 1.

Course 4: has a combination of one tuck and three float stitches, and comes in a sequence of two miss, one tuck, and one miss stitches.

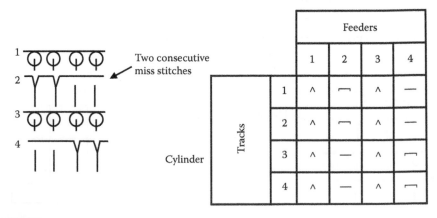

			Feeders			
			1	2	3	4
		1	∧	⌐	∧	—
		2	∧	⌐	∧	—
Tracks		3	∧	—	∧	⌐
		4	∧	—	∧	⌐

FIGURE 9.16
2 × 2 Fleece.

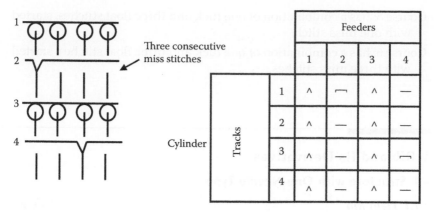

FIGURE 9.17
3 × 1 Fleece.

9.3.3.4 Double Fleece

The design has four courses repeat. The first and third courses represent the ground yarn, while the second and fourth courses represent the fleece yarn. The design has all types of stitches involved in this design. The cam arrangement and design are shown in Figure 9.18.

Design Repeat:

Course 1: has all the knit stitches.

Course 2: has a combination of one tuck and three float stitches, started with a tuck stitch.

Course 3: has a combination of one tuck and three float stitches, started with two miss stitches.

Course 4: same as Course 1.

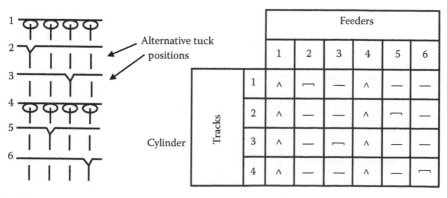

FIGURE 9.18
Double fleece.

Course 5: has a combination of one tuck and three float stitches, started with one miss stitch.

Course 6: has a combination of one tuck and three float stitches, started with three miss stitches.

9.4 Rib and Its Derivatives

9.4.1 Structure with One-Needle Type

9.4.1.1 Plain Rib

The plain rib is a very basic design of rib gating machine. This is formed by all the knit stitches. The design has only one course repeat. This can be produced on one feeder repeat. The cam arrangement and design are shown in Figure 9.19.

Needle arrangements for all the single-needle-type designs of rib fabric are shown in Figure 9.20.

9.4.1.2 Alternative Half Milano

This structure repeat consists of totally four courses. There are two stitch types, knit and miss, involved in this design. A minimum of one feeder is required to produce this design. The cam arrangement and design are shown in Figure 9.21.

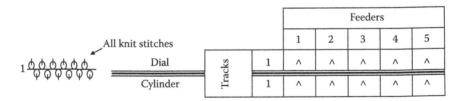

FIGURE 9.19
Pain rib design and its cam arrangement.

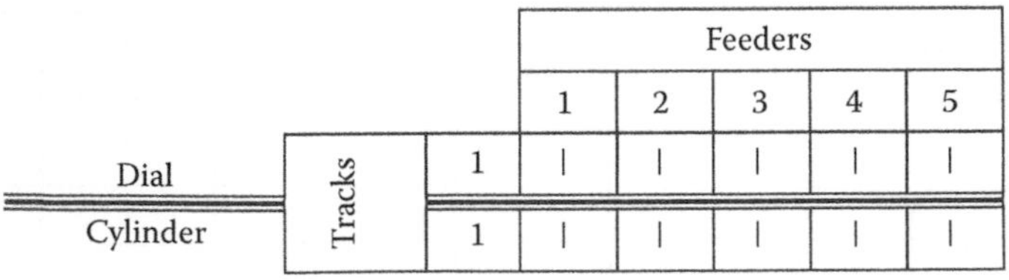

			Feeders				
			1	2	3	4	5
Dial	Tracks	1	|	|	|	|	|
Cylinder		1	|	|	|	|	|

FIGURE 9.20
Needle arrangement for single-needle-type rib designs.

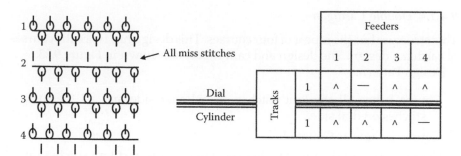

FIGURE 9.21
Alternative half milano.

Stitch Pattern:

Course 1: has all the knit stitches on both sides.

Course 2: has all the knit stitches on cylinder side and all the miss stitches on dial side.

Course 3: same as Course 1.

Course 4: has all the knit stitches on dial side and miss or float stitches on cylinder side. This course has reverse effect of Course 2.

9.4.1.3 Cardigan

The design has only two courses repeat. This design is formed by the knit and tuck stitches. The diagrammatic notation and cam repeat are shown in Figure 9.22.

Design Pattern:

Course 1: has all the knit stitches on cylinder side and all the tuck stitches on dial side.

Course 2: has all the tuck stitches on cylinder side and knit stitches on dial side.

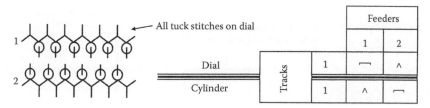

FIGURE 9.22
Cardigan.

9.4.1.4 Double Cardigan

This structure has the repeat of four courses. This design has twice the courses of cardigan design. The design and cam repeat are shown in Figure 9.23.
Design Pattern:

Course 1: has all the knit stitches on cylinder side and all the tuck stitches on dial side.

Course 2: same as Course 1.

Course 3: has all the tuck stitches on cylinder side and knit stitches on dial side.

Course 4: same as Course 3.

9.4.1.5 Double Half Cardigan

This design has extended repeat of half cardigan design. There are four courses repeat in this design. The cam arrangement and design are shown in Figure 9.24.
Design Pattern:

Course 1: has all the knit stitches both on dial and cylinder sides.

Course 2: same as Course 1.

Course 3: has knit stitches on cylinder side and tuck stitches on dial side.

Course 4: same as Course 3.

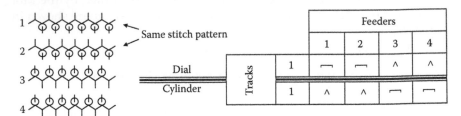

	Feeders			
	1	2	3	4
Dial — Tracks — 1	—	—	∧	∧
Cylinder — 1	∧	∧	—	—

FIGURE 9.23
Double cardigan.

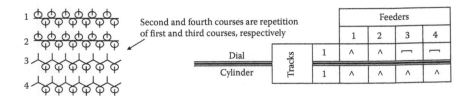

	Feeders			
	1	2	3	4
Dial — Tracks — 1	∧	∧	—	—
Cylinder — 1	∧	∧	∧	∧

FIGURE 9.24
Double half cardigan.

9.4.1.6 Half Cardigan

This pattern consists of two courses. This design is formed by a combination of knit and tuck stitches. The cam arrangement and design are shown in Figure 9.25.
Design Pattern:

Course 1: has all the knit stitches.

Course 2: has all the knit stitches on cylinder side and tuck stitches on dial side.

9.4.1.7 Half Cardigan Double Sided

This design has formed by a little variation in cardigan design. The design has four courses. First and third courses are the same. The second and fourth courses are formed by the alternate positions of tuck and knit stitches. The cam arrangement and design are shown in Figure 9.26.
Design Pattern:

Course 1: has all the knit stitches on both dial and cylinder sides.

Course 2: has all the knit stitches on cylinder side and tuck stitches on dial side.

Course 3: same as Course 1.

Course 4: has all the knit stitches on dial side and tuck stitches on cylinder side.

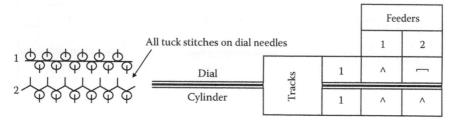

FIGURE 9.25
Half cardigan.

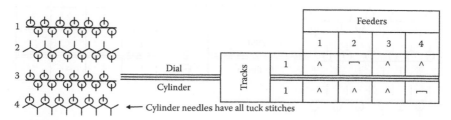

FIGURE 9.26
Half cardigan double sided.

9.4.1.8 Half Milano

This design has a repeat of two courses or feeders. This design is formed by the knit and miss stitches. The cam arrangement and design are shown in Figure 9.27.
Design Pattern:

Course 1: has all the knit stitches.

Course 2: has all the knit stitches on cylinder side and miss or float stitches on dial side.

9.4.1.9 Milano Rib

This is an extended design of half milano design. The additional one course is added in half milano design; thus, the design has totally three courses. The cam arrangement and design are shown in Figure 9.28.
Design Pattern:

Course 1: has all the knit stitches both on dial and on cylinder sides.

Course 2: has all the knit stitches on dial side and miss stitches on cylinder side.

Course 3: has all the knit stitches on cylinder side and miss or float stitches on dial side.

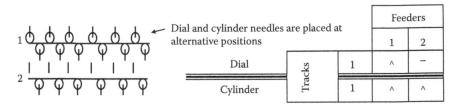

FIGURE 9.27
Half milano.

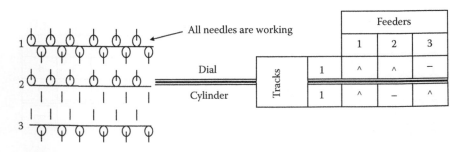

FIGURE 9.28
Milano rib.

9.4.1.10 Rib Ripple

This design is formed by the repeat that consists of three courses. There are two stitch types in this design. Courses 2 and 3 are the same. The cam arrangement and design are shown in Figure 9.29.

Design Pattern:

Course 1: has all the knit stitches both on dial and on cylinder sides.

Course 2: has all the knit stitches on cylinder side and miss or float stitches on dial side.

Course 3: same as Course 2.

9.4.1.11 Ripple Cardigan

This design has a total four courses repeat. The last three courses are the same and have a tuck and knit stitch effect on both dial and cylinder. The cam arrangement and design are shown in Figure 9.30.

Design Pattern:

Course 1: has all the knit stitches on both dial and cylinder sides.

Course 2: has all the tuck stitches on dial side and knit stitches on cylinder side.

Course 3: same as Course 2.

Course 4: same as Course 2.

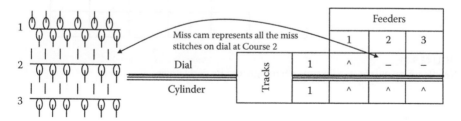

FIGURE 9.29
Rib ripple.

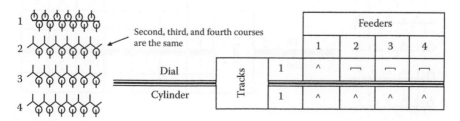

FIGURE 9.30
Ripple cardigan.

9.4.2 Structure with Two-Needle Type

The derivatives of rib design that have two different types of stitches in the same course comes under this category. This design requires two tracks or needle types to produce on machine. The cams are placed according to the type of stitches in the respective course. The general configuration for needle in these structures is given in needle repeat. On the cylinder side, the first needle of Track 1 and the second needle of Track 2 are placed. The same pattern is followed in dial side. The needle repeats are shown in Figure 9.31. A few structures of the rib two-needle type are given below.

9.4.2.1 Belgian Double Pique

This design has a repeat of six courses. The design is formed by the knit and miss stitches. The cam arrangement and design are shown in Figure 9.32.

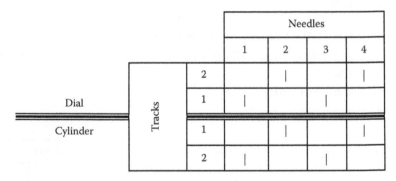

		Needles			
		1	2	3	4
Dial	2		|		|
Dial	1	|		|	
Cylinder	1		|		|
Cylinder	2	|		|	

FIGURE 9.31
Needle repeat of two-needle-type rib design and its derivatives.

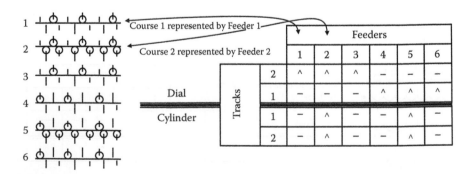

1 ...
Course 1 represented by Feeder 1

2 ...
Course 2 represented by Feeder 2

3 ...
4 ...
5 ...
6 ...

		Feeders					
		1	2	3	4	5	6
Dial	2	^	^	^	–	–	–
Dial	1	–	–	–	^	^	^
Cylinder	1	–	^	–	–	^	–
Cylinder	2	–	^	–	–	^	–

FIGURE 9.32
Belgian double pique.

Design Pattern:

Course 1: has all the miss or float stitches on cylinder side, and dial side has knit stitches on even needles.

Course 2: has all the knit stitches on cylinder side, and dial side has knit stitches on even needles only.

Course 3: same as Course 1.

Course 4: has all the miss or float stitches on cylinder side, and dial side has knit stitches on odd needles.

Course 5: has all the knit stitches on cylinder side, and dial side has knit stitches on odd needles only.

Course 6: same as Course 4.

9.4.2.2 Dutch Double Pique

This design is formed by four feeders. The design is based on knit and miss stitches. The cam arrangement and design are shown in Figure 9.33.

Design Pattern:

Course 1: has all the miss or float stitches on cylinder side, and dial side has knit stitches on even needles.

Course 2: has all the knit stitches on cylinder side, and dial side has knit stitches on odd needles only.

Course 3: has all the miss or float stitches on dial side, and cylinder side has knit stitches on even needles only.

Course 4: has all the knit stitches on dial side, and cylinder side has knit stitches on odd needles only.

9.4.2.3 Fillet

This design has a total of six courses. This design has all the three different types of stitches. The cam arrangement and design are shown in Figure 9.34.

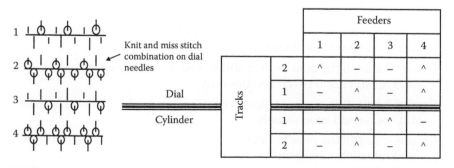

		Feeders			
		1	2	3	4
Tracks	2	^	–	–	^
	1	–	^	–	^
	1	–	^	^	–
	2	–	^	–	^

FIGURE 9.33
Dutch double pique.

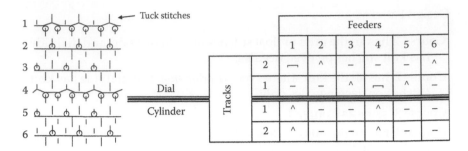

FIGURE 9.34
Fillet.

Design Pattern:

Course 1: has all the knit stitches on cylinder side, and dial side has tuck stitches on even needles.

Course 2: has all the miss or float stitches on cylinder side, and dial side has knit stitches on even needle only.

Course 3: has all the miss or float stitches on cylinder side, and dial side has knit stitches on odd needles.

Course 4: has all the knit stitches on cylinder side, and dial side has tuck stitches on odd needles.

Course 5: same as Course 3.

Course 6: same as Course 2.

9.4.2.4 Flemish Double Pique

The design has a total of eight courses. The design is formed by the knit and miss stitches at different stitching positions of dial and cylinder sides. The cam arrangement and design are shown in Figure 9.35.

Deign Pattern:

Course 1: has all the knit stitches on cylinder side, and dial side has knit stitches on even needles only.

Course 2: has all the miss or float stitches on dial side, and cylinder side has knit stitches on even needles only.

Course 3: has all the knit stitches on dial side, and cylinder side has knit stitches on odd needles only.

Course 4: has all the miss or float stitches on cylinder side, and dial side has knit stitches on even needles only.

Course 5: has all the knit stitches on cylinder side, and dial side has knit stitches on odd needles only.

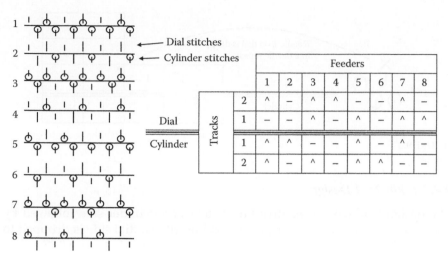

FIGURE 9.35
Flemish double pique.

Course 6: has all the miss or float stitches on dial side, and cylinder side has knit stitches on odd needles only.

Course 7: has all the knit stitches on dial side, and cylinder side has knit stitches on even needles only.

Course 8: has all the miss or float stitches on cylinder side, and dial side has knit stitches on odd needles only.

9.4.3 Drop Needle Design

The drop needle design on circular rib machine can be produced by removing the selective needle from dial and cylinder sides. The most common rib drop needle designs are Rib 2 × 1, Rib 2 × 2, Rib 3 × 2, etc. These designs give different esthetics and functionalities to the fabric. The drop needle designs are produced by placing all the knit cams in a single track both on dial and on cylinder sides. The cam repeat is shown in Figure 9.36. The different drop needle effects are produced by the needle repeat.

			Feeders				
			1	2	3	4	5
Dial	Tracks	1	^	^	^	^	^
Cylinder		1	^	^	^	^	^

FIGURE 9.36
Cam repeat for rib drop needle design.

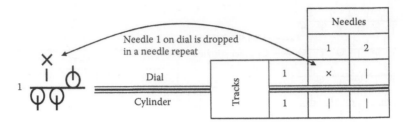

FIGURE 9.37
Rib 2 × 1.

9.4.3.1 Rib 2 × 1 Design

The design has two consecutive knit stitches on cylinder side followed by one knit stitch on dial side. The every odd needle on dial side is dropped to get this effect. The needle repeat and design are shown in Figure 9.37.

9.4.3.2 Rib 2 × 2 Design

The design has two knit stitches on cylinder side and two stitches on dial side in a repeat. The third needle is dropped on cylinder and very first needle is dropped on dial in a needle repeat. The needle repeat and design are shown in Figure 9.38.

9.4.3.3 Rib 3 × 2 Design

The design has three stitches on cylinder side and two stitches on dial side. One needle dropped on cylinder and two needle dropped on dial in a needle repeat. The needle repeat and design are shown in Figure 9.39.

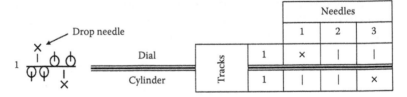

FIGURE 9.38
Rib 2 × 2.

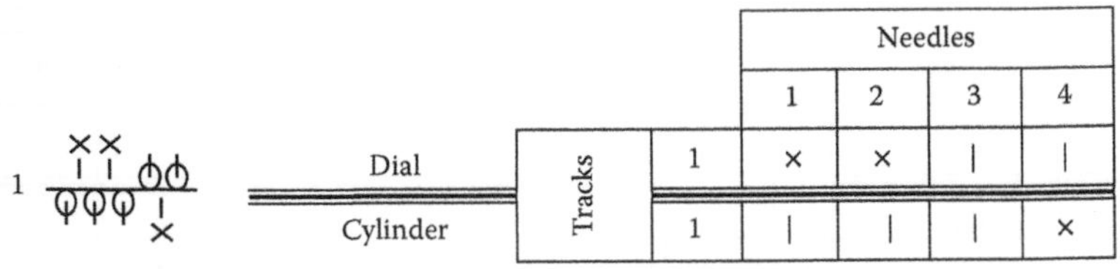

FIGURE 9.39
Rib 3 × 2.

9.5 Interlock and Its Derivatives

9.5.1 Structure with Two-Needle Type

The interlock structures require a minimum of two tracks to produce on cam track system because the needles are placed opposite to each other on dial and cylinder sides. The two types of stitches are used to control the needles on interlock gating machines. The needles pattern is shown in Figure 9.40. The combination of long and short needles are used on interlock machine. Track 1 represents the short needle, while Track 2 represents the long needle.

9.5.1.1 Plain Interlock

The plain interlock structure has two feeders repeat. The plain interlock structure is formed by all the knit stitches. The needle repeat and design are shown in Figure 9.41.

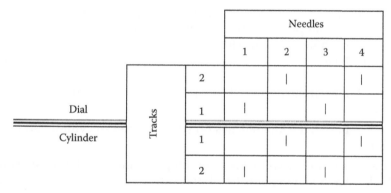

FIGURE 9.40
Needle repeat of plain interlock design.

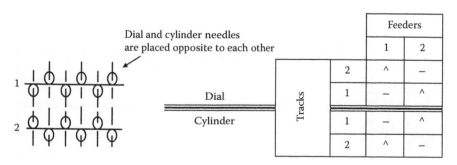

FIGURE 9.41
Design and cam repeat of plain interlock.

Design Pattern:

Course 1: has all the knit stitches on short needles.
Course 2: has all the knit stitches on long needles.

9.5.1.2 Interlock Cross Tubular

The design has two feeders repeat. In this design, the needles are arranged in 2 × 2 manner. Two short and two long needles are placed in a sequence. The needle repeat and design are shown in Figure 9.42.
Design Pattern:

Course 1: has all the knit stitches on short needles.
Course 2: has all the knit stitches on long needles.

9.5.1.3 Interlock Half Cardigan

This design has four feeders repeat. The design has knit, tuck, and miss stitches at alternate positions of dial and cylinder side. The needle repeat and design are shown in Figure 9.43.
Design Pattern:

Course 1: has all the knit stitches on long needles of cylinder and dial sides.
Course 2: has all the knit stitches on short needles of cylinder and dial sides.
Course 3: has all the knit stitches on long needles of cylinder side and tuck stitches on long needles of dial side.
Course 4: has all the knit stitches on short needles of cylinder side and tuck stitches on short needles of dial side.

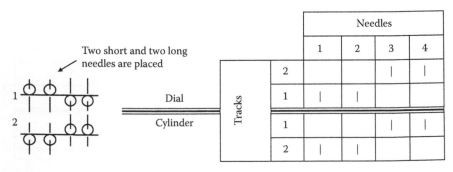

FIGURE 9.42
Design and needle repeat of interlock cross tubular.

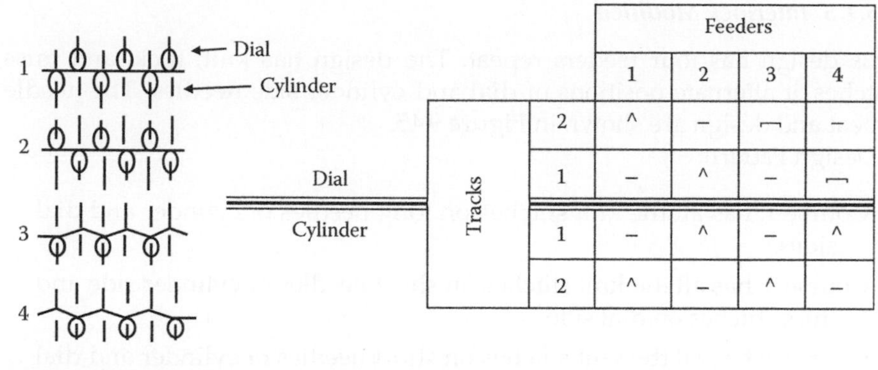

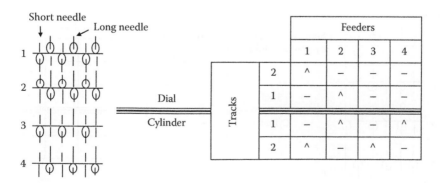

Figure 9.43 illustration: needle repeat (rows 1–4) with Dial and Cylinder labels, and cam repeat table.

	Feeders			
	1	2	3	4
Dial 2	∧	–	⌐	–
Dial 1	–	∧	–	⌐
Cylinder 1	–	∧	–	∧
Cylinder 2	∧	–	∧	–

FIGURE 9.43
Cam repeat and design of interlock half cardigan.

9.5.1.4 Interlock Half Milano

This design has four feeders repeat. The design has knit and miss stitches at alternate positions of dial and cylinder side needles. The needle repeat and design are shown in Figure 9.44.

Design Pattern:

Course 1: has all the knit stitches on long needles of cylinder and dial sides.

Course 2: has all the miss stitches on short needles of cylinder and dial sides.

Course 3: has all the knit stitches on long needles of cylinder side and miss stitches on dial side.

Course 4: has all the knit stitches on short needles of cylinder side and miss stitches on dial side.

Figure 9.44 illustration: needle repeat (rows 1–4) with Short needle and Long needle labels, Dial and Cylinder labels, and cam repeat table.

	Feeders			
	1	2	3	4
Dial 2	∧	–	–	–
Dial 1	–	∧	–	–
Cylinder 1	–	∧	–	∧
Cylinder 2	∧	–	∧	–

FIGURE 9.44
Cam repeat and design of interlock half milano.

9.5.1.5 Interlock Modified

This design has four feeders repeat. The design has knit, tuck, and miss stitches at alternate positions of dial and cylinder side needles. The needle repeat and design are shown in Figure 9.45.
    Design Pattern:

Course 1: has all the knit stitches on long needles of cylinder and dial sides.

Course 2: has all the knit stitches on short needles of cylinder side and miss stitches on dial side.

Course 3: has all the knit stitches on short needles of cylinder and dial sides.

Course 4: has all the knit stitches on long needles of cylinder side and miss stitches on dial side.

9.5.1.6 Interlock Cross Miss

This design has four feeders repeat. The design has knit, tuck, and miss stitches at alternate positions of dial and cylinder side needles. The needle repeat and design are shown in Figure 9.46.
    Design Pattern:

Course 1: has all the knit stitches on long needles of cylinder and dial sides.

Course 2: has all the knit stitches on short needles of cylinder and dial sides.

Course 3: has all the knit stitches on long needles of cylinder side and miss stitches on dial side.

Course 4: same as Course 2.

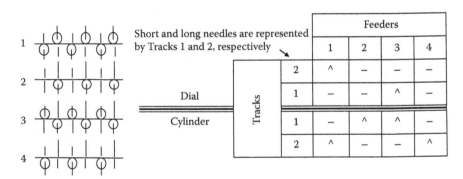

Short and long needles are represented by Tracks 1 and 2, respectively

		Feeders			
	Tracks	1	2	3	4
Dial	2	^	–	–	–
	1	–	–	^	–
Cylinder	1	–	^	^	–
	2	^	–	–	^

FIGURE 9.45
Cam repeat and design of interlock modified.

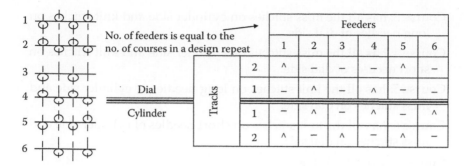

	Feeders					
Tracks	1	2	3	4	5	6
2	^	–	–	–	^	–
1	–	^	–	^	–	–
1	–	^	–	^	–	^
2	^	–	^	–	^	–

FIGURE 9.46
Cam repeat and design of interlock cross miss.

Course 5: same as Course 1.

Course 6: has all the knit stitches on short needles of cylinder side and miss stitches on dial side.

9.5.1.7 *Interlock Cross Relief*

This design has eight feeders repeat. The design has knit and miss stitches at different positions of dial and cylinder side needles. The needle repeat and design are shown in Figure 9.47.
Design Pattern:

Course 1: has all the knit stitches on long needles of cylinder and dial sides.

Course 2: has all the knit stitches on short needles of cylinder and dial sides.

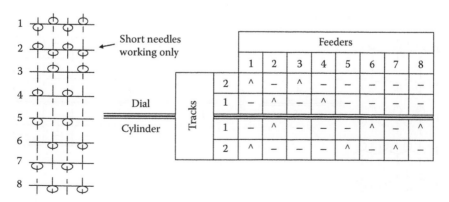

	Feeders							
Tracks	1	2	3	4	5	6	7	8
2	^	–	^	–	–	–	–	–
1	–	^	–	^	–	–	–	–
1	–	^	–	–	–	^	–	^
2	^	–	–	–	^	–	^	–

FIGURE 9.47
Cam repeat and design interlock cross relief.

Course 3: has all the miss stitches on cylinder side and knit stitches on long needles of dial side.

Course 4: has all the miss stitches on cylinder side and knit stitches on short needles of dial side.

Course 5: has all the knit stitches on long needles of cylinder side and miss stitches on dial side.

Course 6: has all the knit stitches on short needles of cylinder side and miss stitches on dial side.

Course 7: same as Course 5.

Course 8: same as Course 6.

9.5.1.8 Interlock Double Tuck

This design has six feeders repeat. The design has knit, tuck, and miss stitches at alternate positions of dial and cylinder side needles. The needle repeat and design are shown in Figure 9.48.
Design Pattern:

Course 1: has all the knit stitches on long needles of cylinder and dial sides.

Course 2: has all the tuck stitches on short needles of cylinder side and knit stitches on short needles of dial side.

Course 3: has all the knit stitches on long needles of cylinder side and tuck stitches on long needles of dial side.

Course 4: has all the knit stitches on short needles of cylinder and dial sides.

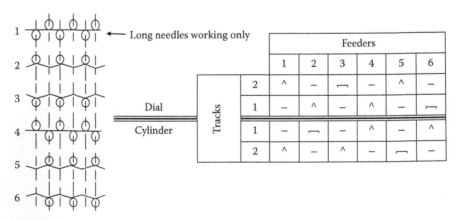

FIGURE 9.48
Cam repeat and design of interlock double tuck.

Course 5: has all the tuck stitches on long needles of cylinder side and knit stitches on long needles of dial side.

Course 6: has all the knit stitches on short needles of cylinder side and tuck stitches on short needles of dial side.

9.5.1.9 Interlock Piquette

This design has six feeders repeat. The design has knit and miss stitches at alternate positions of dial and cylinder side needles. The needle repeat and design are shown in Figure 9.49.

Design Pattern:

Course 1: has all the knit stitches on long needles of cylinder and dial sides.

Course 2: has all the miss stitches on long needles of cylinder side and knit stitches on long needles of dial side.

Course 3: has all the knit stitches on short needles of cylinder side and miss stitches on dial side.

Course 4: has all the knit stitches on short needles of cylinder and dial sides.

Course 5: has all the miss stitches on cylinder side and knit stitches on short needles of dial side.

Course 6: has all the knit stitches on long needles of cylinder side and miss stitches on dial side.

9.5.1.10 Interlock Rodier

This design has six feeders repeat. The design has knit, tuck, and miss stitches at alternate positions of dial and cylinder side needles. The needle repeat and design are shown in Figure 9.50.

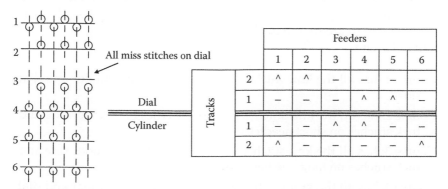

FIGURE 9.49
Cam repeat and design of interlock piquette.

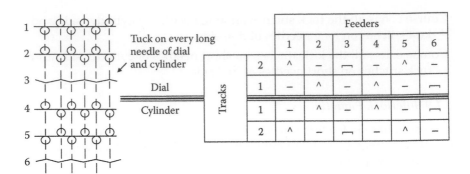

FIGURE 9.50
Cam repeat and design of interlock rodier.

Design Pattern:

Course 1: has all the knit stitches on long needles of cylinder and dial sides.

Course 2: has all the knit stitches on short needles of cylinder and dial sides.

Course 3: has all the tuck stitches on long needles of cylinder and dial sides.

Course 4: same as Course 2.

Course 5: same as Course 1.

Course 6: has all the tuck stitches on short needles of cylinder and dial sides.

9.5.1.11 Pin Tuck

This design has six feeders repeat. The design has knit, tuck, and miss stitches at alternate positions of dial and cylinder side needles. The needle repeat and design are shown in Figure 9.51.

Design Pattern:

Course 1: has all the knit stitches on long needles of cylinder and dial sides.

Course 2: has all the knit stitches on short needles of cylinder side and tuck stitches on short needles of dial side.

Course 3: has all the knit stitches on long needles of cylinder side and tuck stitches on long needles of dial side.

Course 4: has all the knit stitches on short needles of cylinder and dial sides.

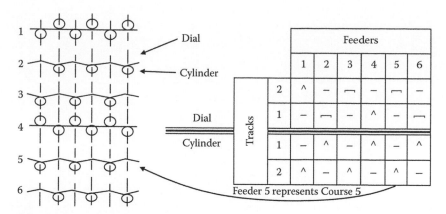

FIGURE 9.51
Cam repeat and design of pin tuck.

Course 5: has all the knit stitches on long needles of cylinder side and tuck stitches on long needles of dial side.

Course 6: has all the knit stitches on short needles of cylinder side and tuck stitches on short needles of dial side.

Bibliography

Ashraf, W., Nawab, Y., Maqsood, M., Khan, H., and Awais, H. 2015. Development of seersucker knitted fabric for better comfort properties and aesthetic appearance. *Fiber and Polymer* 16(3): 699–701. doi:10.1007/s12221-015-0699-0.

Anbumani, N. 2005. *Knitting Fundamentals, Machine, Structure and Development*. New Delhi: New Age International(P) Limited.

Iyer, C., Mammel, B., and Schäch, W. 1992. *Circular Knitting*. Bamberg: Meisenbach GmbH.

Nawab, Y. ed. 2016. *Textile Engineering: An Introduction*. 1st ed. De Gruyter, Berlin/ Boston, Germany.

Ray, S. C. 2012. *Fundamentals and Advances in Knitting Technology. Zhurnal Eksperimental'noi I Teoreticheskoi Fiziki*. New Delhi: Woodhead Publishing India.

10

Patterning in Warp Knitting

Muzammal Hussain and Yasir Nawab

CONTENTS

10.1 Knitting Elements of Warp Knitting Machine

In warp knitting machine, unlike weft knitting machine, yarn comes from warp beams. Individual yarns are wrapped on beam. Each needle is fed individually, and sheet of knitted fabric is formed, which is then wound on take-down roller. Both weft knitted and warp knitted fabrics are formed by interlooping, yet there is difference between them in the structure and loop formation process. Unlike weft knitted fabric, warp knitted fabric is formed in lengthwise direction. Knitting elements used in warp knitting are also different from those used in weft knitting.

10.1.1 Guides and Guide Bar

Warp yarn passes through guides. Guide is a thin metal plate having a hole at one end through which warp yarn passes. Guides are placed on a lead piece of 1 inch. At one end of guide, there is a hole, and the other end is attached with a lead piece as shown in Figure 10.1. This lead piece is placed on a bar extended across the full width of machine. Gauge of guide bar is same as that of machine. The number of guide bars on a machine is equal to the number of warp beams on the machine. Each guide bar is threaded by yarns of individual warp beams. In warp knitting, unlike weft knitting, needle does not take yarn and knit, but guide bar wraps the yarn around the hook of needle through lapping movement. For commercially acceptable structure, at least two guide bars are used. These guide bars are given identical lapping movement, so consumption of yarn will be the same. Different guide bars can have different lapping movement, so consumption of yarn will also be different. Guides are placed in such a way that there is needle between two guide bars. Needle should be exactly in between two guides; otherwise, it will hit the needle during lapping movement (Ray, 2011).

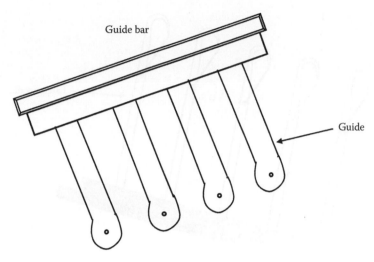

FIGURE 10.1
Guide and guide bar.

10.1.2 Needle and Needle Bar

All three types of needles are used in warp knitting machine. Spring beard needle is used for tricot knitting machine as shown in Figure 10.2; for raschel knitting machine, latch needle is used as shown in Figure 10.3; and for both tricot and raschel knitting machines, compound needle can be used. Needles are placed in tricks of bar according to the gauge of machine. This bar is

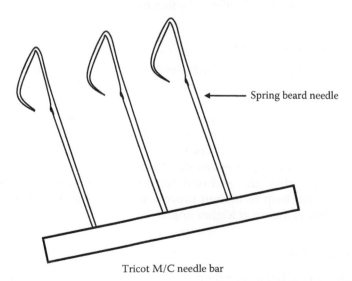

FIGURE 10.2
Needle bar of tricot m/c.

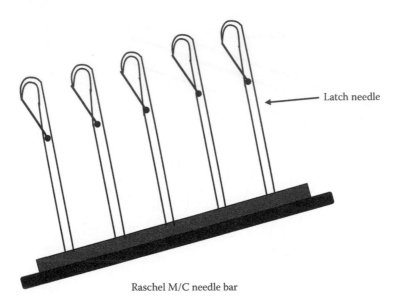

Latch needle

Raschel M/C needle bar

FIGURE 10.3
Needle bar of raschel m/c.

called needle bar. In weft knitting, needles move in track one by one according to design, while in warp knitting machine, all needles are given identical upward and downward motions, as in single course, all needles have identical movement, so needles attached on a needle bar are lifted up and down with the help of cam fitted outside the machine.

10.1.3 Sinker and Sinker Bar

Sinker is a thin metallic plate. Each needle is placed between two sinkers. Sinkers are fitted on a bar through a lead piece. All sinkers are given identical movement through shaft. Sinkers move forward and backward. On tricot and raschel knitting machines, sinkers have different functions. Fabric is held down with the help of neb and throat, while belly portion is used for knocking over of loop. In tricot knitting machine, sinkers are involved in whole loop formation process as shown in Figure 10.4, whereas in raschel knitting machine, they are used only for holding down purpose as shown in Figure 10.5 and loop formation process is assisted with high take-down tension as fabric is drawn at higher angle.

10.1.4 Pressor Bar

When using spring beard needle, there is a need of some external elements to close the hook of the needle during casting off. As spring beard needle is used on tricot warp knitting machine, an additional element, called pressor

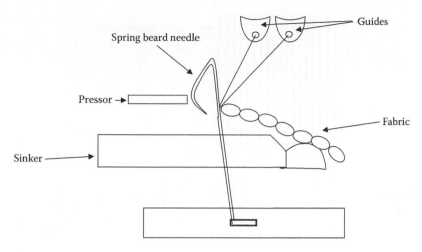

FIGURE 10.4
Loop formation process of tricot m/c.

bar, is required across the full width of machine to close the hook. This bar works when needles start descending after receiving new yarn and helps in casting off of old loop.

10.1.5 Latch Wire

During loop formation on warp raschel knitting, when loop clears the latch and comes on stem of needle, sometime there is chance of flicking back of

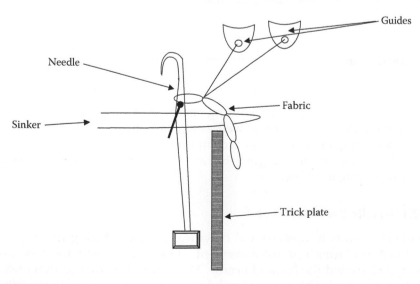

FIGURE 10.5
Loop formation process of raschel m/c.

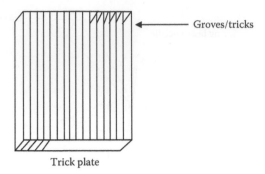

Trick plate

FIGURE 10.6
Trick plate.

latch. This will cause closure of needle and prevent from taking new yarn. To prevent latch to close the hook, a metallic wire extended across the full width of needle bed is used.

10.1.6 Trick Plate

This is a vertical plate extended across the full width of machine, which is also sometimes called "knock over comb." The latch needle moves up and down between the blades or verges of trick. The top edge of trick plate acts as the firm surface that assists in knock over. The blades of trick plates ensure exact knocking over of stitches especially in longer underlapping (Ajgaonkar 1998). Trick plate is shown in Figure 10.6.

10.2 Movement of Different Warp Knitting Elements

For fabric formation on warp knitting machine, there is need to feed the yarn to the individual needle as well as to connect the adjacent wales, and the guides of a guide bar execute a compound lapping movement. Motions of needle bar, sinker bar, and guide bar are synchronized for loop formation on warp knitting machine.

10.2.1 Needle Bar Movement

Needle is given only upward and downward motion during the loop formation cycle. During upward movement, old loop is cleared and new yarn is wrapped around the hook of needle. When needle is moved downward, knock over is done. This upward and downward movement of the needle is given through an eccentric shaft which is attached to needle bar.

10.2.2 Guide Bar Movement

Guide bar performs swinging and shogging motions as shown in Figure 10.7. The purpose of these motions is to feed the yarn to needle and connect the adjacent wales during loop formation. Shogging and swinging in combination are called lapping movement. These two motions work at right angle to each other. Underlap and overlap are the result of this lapping movement. Swinging motion is either from front to back or from back to front side of machine. In contrast, shogging movement is either from left to right or right to left side of machine. Swinging motion forms the two limbs of loop, and shogging motion causes the overlap and underlap. Extent of overlap and underlap depends on the lateral movement given by pattern chain or pattern wheel to guide bar. There is separate pattern chain and pattern wheel for each guide bar. The extent and motion of each guide bar are controlled individually, which decide the type of fabric formed. Overlap is at the front side of the needle and underlap is at the back side of needle as shown in Figure 10.8.

10.2.3 Sinker Bar Movement

There is one needle between every two sinkers. Sinkers are supportive elements in the loop formation process. Sinkers move to and fro during lapping

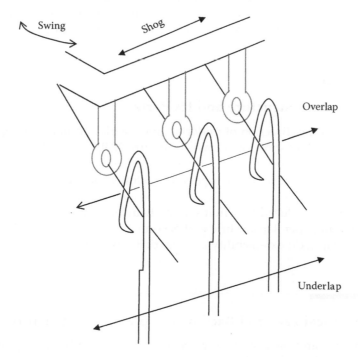

FIGURE 10.7
Guide bar movement.

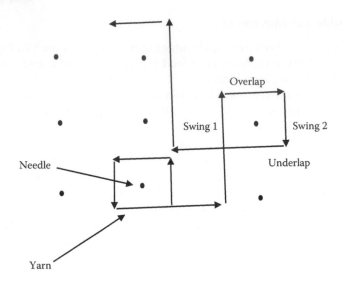

FIGURE 10.8
Overlap and underlap movement of guide bar.

movement, and this forward and backward movement is given to sinkers through sinker bar which get motion from eccentric shaft. Sinkers remain at forward position during whole loop formation cycle, and as soon as needle descends after receiving new yarn, sinkers move backward to assist casting off.

10.3 Warp Knit Structure and Its Parts

Depending on the direction of underlap and overlap in machine, open and closed laps can be formed as shown in Figure 10.9. A closed lap is formed when an underlap and an overlap follow in opposite direction to each other. An open lap is formed when underlap follows the overlap in the same direction.

A fabric having closed laps in it will be heavier and thicker as compared to fabric having open laps. Fabric will become more compact and less extensible. Open laps add extensibility and elasticity to fabric.

10.4 Technical Face and Back of Warp Knitted Structure

Technical face and back sides of warp knit fabrics can be distinguished. A technical face side is loop side or side having overlap prominent. In contrast, technical back side has float or underlap prominent. For double needle bar

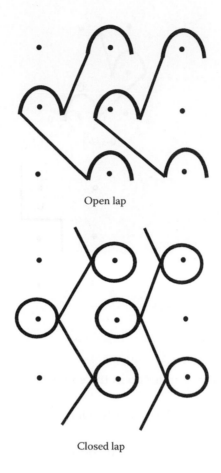

Open lap

Closed lap

FIGURE 10.9
Open and closed lap.

structures, floats or under-laps are entrapped between two layers of loops or overlaps, so both sides are taken as technical face sides (AU 2011).

10.5 Variations in Underlap/Overlap

There are five variations possible in overlap and underlap as shown in Figure 10.10. These variations are possible due to one or more of the following lapping movements of guide bar:

1. If an overlap follows an underlap in its opposite direction, a closed lap is formed.

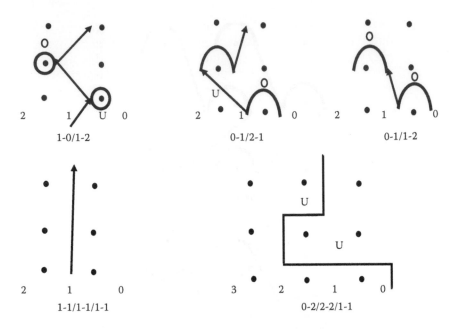

FIGURE 10.10
Lapping motions.

2. If an overlap follows underlap in the same direction, an open lap is formed.

3. When there is no underlap but only overlap is present, an open lap is formed.

4. When there is only underlap but no overlap is present, a laid-in structure is formed.

5. When there is neither overlap nor underlap, a miss lap is formed.

The last two movements do not exist alone, so there is need of another guide bar lap to hold them (Spencer 2001) (Ray, 2011).

10.6 Effect of Underlap and Overlap on Fabric Properties

Shogging motion produces overlap and underlap in fabric, and usually, the overlap shog is across a single needle. On conventional warp knitting machine, a needle receives yarn from at least one guide in a knitting cycle. If needle fails to receive yarn, then there will be a hole in the fabric. An overlap shog is usually across one needle, but it may also be across two needles, which produces severe tension in knitted fabric. Appearance of fabric having two needle overlaps is poor.

The length of underlap is taken as needle space. Length of underlap can vary as per structure requirement. It can be 0–4 needles space. The extent of underlap increases the production speed, but the efficiency of knitting decreases. As underlap is increased, fabric balance shifts from length to width direction. A longer underlap increases the widthwise stability of fabric, and shorter underlap reduces the widthwise stability but increases the lengthwise stability. The length of underlap also affects the weight and thickness of fabric. Fabric weight and thickness increase as more yarn is to fed for longer underlap and longer underlap cover more wales on its way.

10.7 Run-in, Rack, and Its Effect on Fabric

Whenever weft knitted fabric is in consideration, the length of yarn knitted in a single course by all the needles as per design is referred as "course length." In warp knitting term, "run-in per rack" is used instead of course length. Run-in is consumption of yarn in a rack, while rack is a unit of 480 courses. Run-in is measured in millimeters or inches. Multiple guide bars has usually different yarn consumption during fabric formation, therefore, yarn consumption for each guide bar is required to be calculated. For a given machine with a given warp, a longer run-in produces bigger stitches and generally a slacker, looser fabric, whereas a shorter run-in produces smaller and tighter stitches (Spencer 2001).

10.8 Pattering Mechanism in Warp Knitting Machine

As discussed earlier, guide bar is given a shogging movement which is responsible for overlap and underlap. Guide bar moves away from pattering mechanism when shog positively, but it moves toward pattering mechanism when shog negatively. Guide bar can be pushed one, two, three, or more needles as per requirement. If there is no shog, guide bar will remain at same height. The extent of shogging is decided by one of the following pattern controlling mechanisms:

1. Pattern wheel
2. Pattern chain
3. Electronic jacquard

10.8.1 Pattern Wheel

Pattern wheel mechanism is the simplest of all, and the profile of pattern wheel is made of slopes of different heights as shown in Figure 10.11. A bowl works on these slopes, which is connected to a push rod, and the push rod shogs the guide left or right for underlap and overlap. Guide bar is pushed with push rod and reversed back by a reversing spring as shown in Figure 10.12. The pattern wheel is just like cam on which slopes are made through careful cutting of a steel disc. Slopes are made very carefully and precisely so that no disrupted shogging is produced when bowl works on these slopes. The height of slopes decides the amount of lateral shogging. Pattern wheel type mechanism offers the highest speed due to less complexity of design and smoothness of profile.

Pattern wheel is usually divided into 48 segments. If two movements are required for one course, for example, first overlapping and second underlapping, then there will be a total of 24(48/2) courses produced in one revolution. If three movements are required for one course, then 16(48/3) courses will be produced with one revolution of pattern wheel. Pattern wheel can only be used if the total number of divisions on pattern wheel is divisible by the number of movements required for one course.

There are some limitations of using pattering wheel such as there are only limited patterns are possible and only one type of design can be used at a time. Pattern wheel can only be used for a machine for which it is made. This pattern mechanism is used for large production orders, and there are no

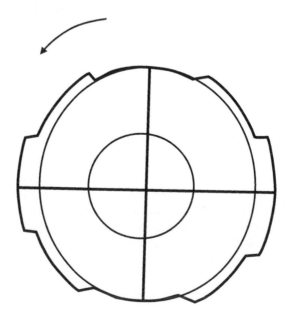

FIGURE 10.11
Pattern wheel.

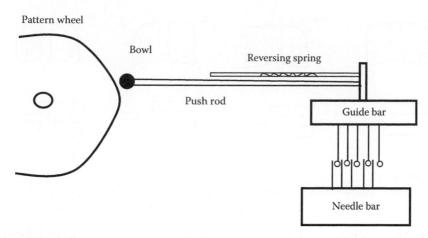

FIGURE 10.12
Pattern wheel working mechanism.

design variations. In spite of all limitations, high speed is possible with less chances of errors.

10.8.2 Pattern Chain

In pattern chain mechanism, chain is constructed by connecting Y-shaped chain links. Chain links are like tuning fork with a tail. Tail of front chain link is placed between forks of rear link. Then, through small pins which pass through, the holes connect the individual links. All the chain links are connected through this way, and profile is formed on which roller of push rod works. Pattern links are made of hardened steel, and very accurate grinding of individual link is done. Too sharp edge will cause an early shog, and a gradual edge will produce a delayed shog (Anbumani 2010).

Four different shapes of chain links are used as shown in Figure 10.13:

Link A is unground.
Link B is ground at front toward fork side.
Link C is ground at back toward tail.
Link D is ground at both front and back.

Different links have different heights, which produce different shogs. Shog is different according to design may be 0, 1, 2, 3, 4 needles, and so on. The chains are placed correctly so that overlap and underlap movements are in sequence, and an overlap should be always toward hook side of needle. The chain links height change as a multiple of machine gauge, as on 28-gauge machine, the height of link will change as multiple of 1/28. Links of each different gauge machine are stamped having gauge and number 0, 1, 2, 3, etc.

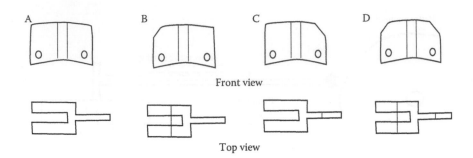

Front view

Top view

FIGURE 10.13
Chain links of pattern chain mechanism.

on it. The link which will put guide bar at the lowest height will be taken as 0. A link of size 1 will move guide bar a 1 needle space and on a 32 gauge machine, the link which will move guide bar 1 needle space will be 1/32 inch thicker than the lowest-height link. As the chain link number increases, the guide bar moves further away from pattern drum. In Figure 10.14, 0 is given to link on which when guide bar will act, it will be on right most position, and when push rod will work on 2 chain link, it will move Guide Bar 2 needle space with respect to 0. Chain link 3 will move 3 needles space, but with respect to chain link 2, it will move 1 needle space. When move to link 2 after 3 will bring one needle back to pattern chain side. These types of links are connected in continuous pattern to make profile of pattern chain, which give shog to guide bar, and the pattern chain mechanism is shown in Figure 10.15 (Ajgaonkar 1998).

10.8.3 Electronic Jacquard

Both pattern chain and pattern wheels are limited and specific in designing whenever a new design is run on machine and then a new pattern chain or

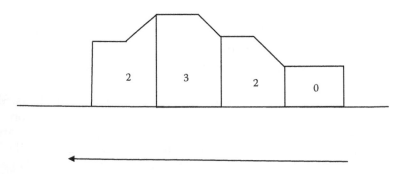

FIGURE 10.14
Profile of pattern chain mechanism.

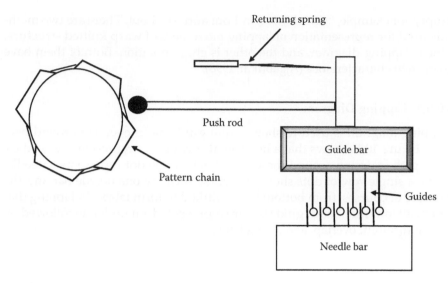

FIGURE 10.15
Pattern chain working mechanism.

pattern wheel is need to be developed. Electronic jacquards are used these days to cope with problems associated with conventional systems. Jacquard system has added more versatility to patterning. Multibar raschel knitting machine is used for lace making, which uses Jacquard system. There are four parts of electronic jacquard system:

Pattern computer
Binary mechanism
Electronically controlled jacquard
Mobile loading device

Most of the jacquard systems use electromagnetic and piezoelectric materials for selection to guide bar (Spencer 2001).

10.9 Warp Knit Structures Representation

There are two main factors of warp knitted structure representation: the first one is the way each guide bar is threaded, and the second one is the lapping movement of each guide bar. Guide bar can be fully threaded or partially threaded; a fully threaded guide bar is represented by a small group of vertical lines "|||||", and if a guide bar is partially threaded, then a vertical line "|" represents that the guide bar is threaded and a "." represents not threaded or

empty, for example, "ǁ·ǁǁ." shows 2 in 1 out and 3 in 1 out. There are two methods used for representation of lapping movement of warp knitted structure. One is lapping diagram, and the other is chain notation. Both of them have their own characteristics (Ajgaonkar 1998).

10.9.1 Lapping Diagram

Lapping diagram is used to show actual guide bar movement around needles. Figure 10.16 shows that a horizontal row of dots represent the needles. Yarn path is drawn around these dots. Each single dot represents a needle, while a single row of dots show the course form by one needle bar and the number of courses, from bottom to top, linked to form fabric. In lapping diagram, all four motions of guide bar are represented, an underlap followed by swing through, overlap, and swing back.

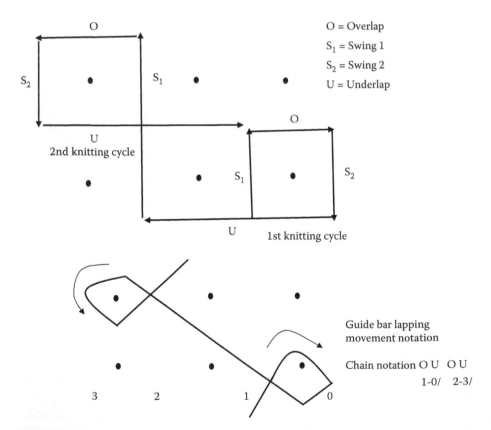

FIGURE 10.16
Lapping movement of guide bar around needle.

10.9.2 Chain Notation

Chain notation represents the correct sequence of chain link arrangement of each guide bar to produce a fabric as shown in Figure 10.16. The number of links for each course is fixed for each machine. A minimum of two links are needed for one course with an underlap joining second link of first course and first link of second course. This is method to write lapping diagram pattern numerically. In this notation, the lowest position, just close to pattern wheel or pattern drum, is given 0 position. Position 1 is given when guide bar shogs 1 needle space relative to the lowest position. Position 2 is given when guide bar shogs further one needle and so on. During chain notation (1-0/2-3/), the first pair of numbers indicates the lapping movement in the first course. The hyphen (-) between numbers indicates the overlap. The difference between two numbers shows the amount of overlap in a course. The pair of numbers also indicates the direction of overlap from 1 to 0. The second pair of numbers indicates the lapping movement in the second course. This pair also indicated the direction of overlap in the second course. The difference between the last number of the first pair and the first number of the second pair shows the amount of underlap. The oblique line "/" shows the underlap movement. Sometimes, two oblique lines "//" are used if chain link is repeating continuously (Ajgaonkar 1998).

10.10 Single Bar Fabrics

Single guide bar warp knit fabrics are not very commonly used due to their less stability, and distortion in the structure which is due to inclining of loops to right or left as loops are not held firmly, these fabrics also have low run resistance, flimsy structure, and less variations in structure design. Single bar structure is not used alone as it is not so stable that it can be used commercially, so it is always used when some other structure is supporting it like pillar stitches (Ajgaonkar 1998).

10.11 Two Guide Bar Structures on Tricot Machine

Warp knitted structures are made with a minimum of two guide bars. A minimum of two guide bar structures are commercially acceptable. Two guide bar structures are produced in such a way that front guide bar dominates both in overlaps and underlaps. Some warp knitted structures produced with double guide bars are discussed below.

10.11.1 Rules of Two Guide Bar Structures

1. When the guide bar swings through for next overlap, between two guide bars front and back, underlap of front guide bar comes on top because back guide underlaps first and front guide bar underlaps later, so it comes on top as shown in Figures 10.17 and 10.18.

2. The front guide bar thread strikes the needle first after overlap during swing back as shown in Figure 10.19. Thread of front guide occupies a lower position during lapping when guide comes in front of machine, so effect of front guide threads is always on lower side, and this is toward technical face side of fabric.

3. A slacker run-in, an open lap, or a short underlap will all tend to cause a warp thread to occupy a low position on the needle. Normally, overlap and underlap of front guide bar come on technical face side of fabric, but this dominance can be overcome if the front and back guide bars work in such a way that front guide bar makes a 2×1 closed lap and back guide bar makes a 1×1 open lap in opposite to each other.

4. When both guide bars moves in opposite direction during overlap. The yarns of both bars twist over each other, so the back bar thread partly shows on top of one side limb. When two bars move in opposite to each other during underlap, this will produce an extra tension

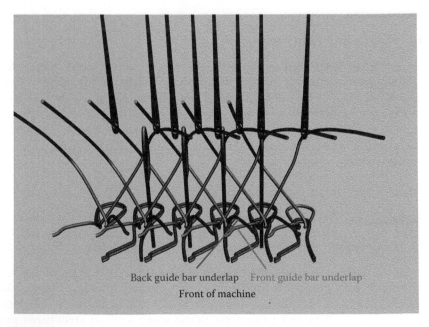

Back guide bar underlap   Front guide bar underlap
Front of machine

FIGURE 10.17
Two guide bar lapping movement.

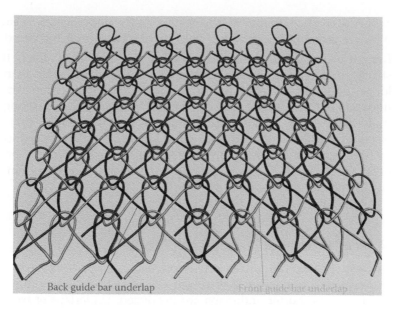

FIGURE 10.18
Technical backside of two guide bar structure.

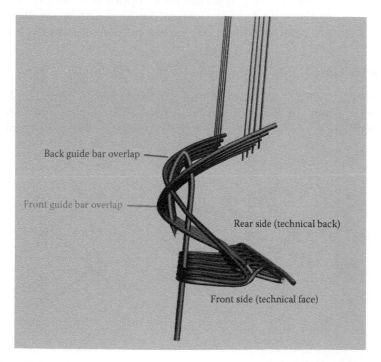

FIGURE 10.19
Front and back guide bar overlap.

at needle head, so overlap will be tighter. As the underlaps cross each other between middle of wales, stability of structures is increased.

5. Length of underlap of front and back guide bars also affects the structure. A short underlap will be at more inclined angle than a longer underlap, and also short underlap is under higher tension. If underlap of front guide bar is shorter and back guide bar has longer underlap, then the front guide bar underlap will hold the back guide bar underlap firmly into structure. If a longer underlap is produced by front guide bar, this will produce a float at face on technical back of fabric. This will increase elasticity of fabric and more freedom of curling on face and back sides of fabric.

10.11.2 Two-Bar Tricot

The simplest structure produced on tricot machine is two-bar tricot fabric often termed as half jersey in America. Two-bar tricot uses a minimum amount of yarn as shorter underlaps are produced. Underlaps of front and back guide bars cross each other at middle of wales. Figure 10.20 shows the lapping movement of front and back guide bars, which is 1-0/1-2// and 1-2/1-0//, respectively. Both of the guide bars move opposite to each other. Cover factor is low for these kinds of structures. These structures split in wales when fine yarn or continuous filament yarn is used during button holing.

10.11.3 Locknit

Locknit structures are the most popular of all knitted structures. Locknit are termed jersey in America and charmeuse in France and Germany. Figure 10.21 shows the lapping movement of two guide bars. The front guide

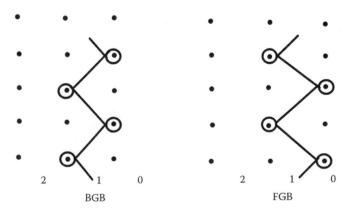

FIGURE 10.20
Lapping movement of tricot structure.

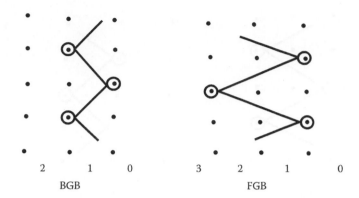

FIGURE 10.21
Lapping movement of locknit structure.

bar makes the lapping movement 1-0/2-3//, and back guide bar makes the lapping movement 1-2/1-0//. Longer overlap of front bar, a maximum of two needles space on back, improves the extensibility, drapability, and handle of fabric. Locknit fabrics are produced usually with more than 37 wales per inch and on 28 gauge machine. Nylon yarn is used, and sometimes, Lycra is also used on back of fabric. Fabric areal density depends on filament denier. Popular areal densities are 32, 82, and 152, which are produced with 20, 40, and 70 denier yarn respectively. Locknit fabrics are elastic that is why used for lingerie and intimate apparel. These fabrics have tendency to curl at edges, so heat setting is used for thermoplastic yarn. Locknit structure has a high shrinkage of usually 20%–30%.

10.11.4 Reverse Locknit

Reverse locknit or reverse jersey is less extensible, and curling effect at edges is also reduced with this structure. Reverse locknit are not as popular as locknit. Figure 10.22 shows the lapping movement of front and back guide bars. Front and back guide bars make 1-2/1-0// and 1-0/2-3// lapping movements, respectively. Lower front guide bar underlap is the reason behind its less curling and shrinkage. A relatively rigid fabric is suitable for shirting as it is more stable than the locknit. It is generally knitted in plain colors.

10.11.5 Sharkskin

Sharkskin is produced by increasing the extent of underlap usually 3–4 needles space. As the extent of underlap is increased, fabric becomes more rigid and less extensible. Hence, properties similar to those of woven fabrics are expected from these fabrics as far as the stability of fabric is concerned. The fabric has high cover and is tight in construction. Application of sharkskin fabric is as print base fabric. The front guide bar makes similar lapping

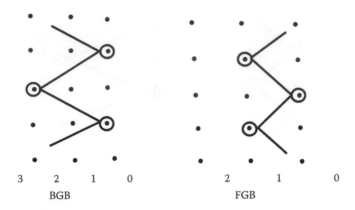

FIGURE 10.22
Lapping movement of reverse locknit structure.

movement as of single-bar tricot structure, the back guide bar makes longer underlaps and the lapping movement is shown in Figure 10.23. Front guide bar movement is 1-0/1-2//, and back guide bar lapping movement is 4-5/1-0//. Figure 10.24 shows back side of fabric.

10.11.6 Queenscord

Queenscord fabrics have even tighter structure than sharkskin due to their distinct characteristics. A front guide bar produces a pillar or chain stitch which holds the fabric very tightly. Pillar stitches are locked very tightly with back bar underlap, so a very low shrinkage of only 1%–6% is possible. Back guide bar makes underlap of 2–3 needles space. Queenscord structure can be printed with several colors, while locknit and other two bar structures are easily distorted, and therefore, proper application of colors with respect to

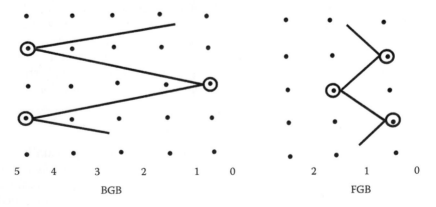

FIGURE 10.23
Lapping movement of sharkskin structure.

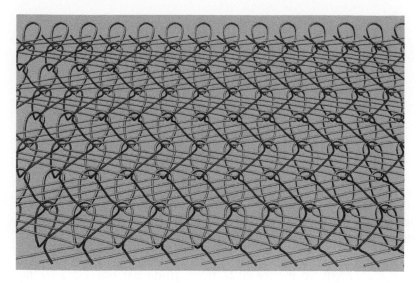

FIGURE 10.24
Reverse side of sharkskin fabric.

each other on such structures becomes difficult. Cord effect is generated in wales direction due to pillar stitches as shown in Figure 10.25. Figures 10.26 and 10.27 show the lapping movement of two different fabrics. In Figure 10.26, front guide bar has 1-2/2-1// lapping and back guide bar has 1-0/3-4//

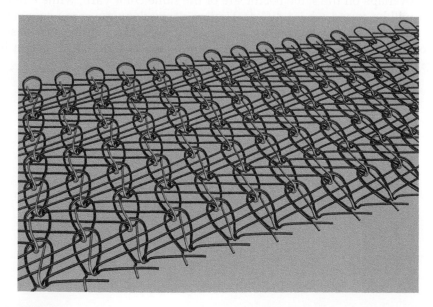

FIGURE 10.25
Queenscord fabric structure.

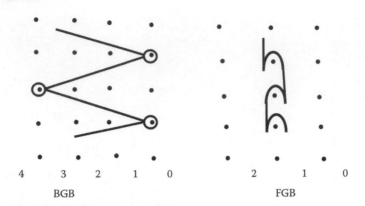

FIGURE 10.26
Queenscord fabric structure with two needles space underlap of back guide bar.

lapping movement. In Figure 10.27, front guide bar has 1-2/2-1// lapping movement and back guide bar has 1-0/2-3// lapping movement.

10.11.7 Double Atlas

Double atlas is produced usually by two guide bars lapping in opposite to each other with identical lapping. Checks, diamonds, and circles are produced with symmetry and balance. An intense area of color is found where both overlaps on the same needle are of the same color yarn, while a fader area will be found where different colors of overlapping yarns on a needle are used. Double atlas fabric has better drapability, handle, and elastic recovery. Construction of atlas can be both open and closed as shown in Figure 10.28.

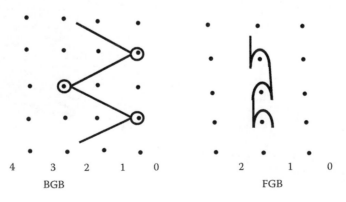

FIGURE 10.27
Queenscord fabric structure with 1 needle space underlap of back guide bar.

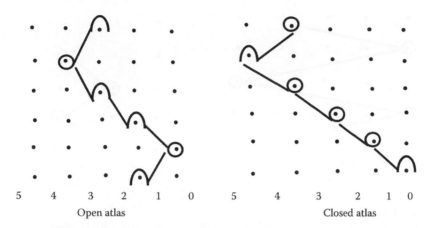

FIGURE 10.28
Lapping movement of open and closed Atlas structure.

10.11.8 Satin

In satin, longer front guide bar underlaps are used. When compared with locknit, these structures have more elasticity. A more reflective surface is produced on back side of fabric when making fabric with continuous filament. Just as in woven satins, the knitted satins have longer floats at back side of the structures, and these floats are actually responsible for higher luster and smoother feel of the fabric when compared to locknit structures made with similar yarn. The technical back side of satin is used as effect side. Floats can be raised to give it fleecy finish. Lapping movement of guide bar can be seen in Figures 10.29 and 10.30. Front guide bar has 3-4/1-0// movement, and back guide bar has 1-0/1-2// movement or 4-5/1-0// of front guide bar and 1-0/1-2// of back guide bar.

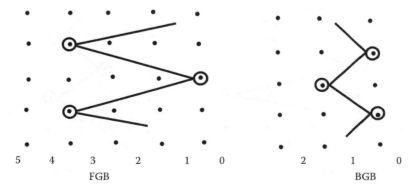

FIGURE 10.29
Lapping movement of Satin Fabric with front guide bar underlap of 2 needle space.

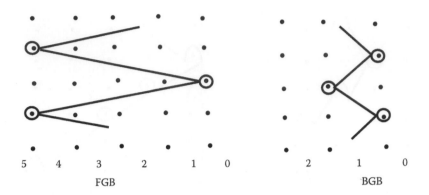

FIGURE 10.30
Lapping movement of Satin Fabric with front guide bar underlap of 3 needle space.

10.11.9 Loop Raised Fabrics

The basic for the construction of these fabrics is similar to that for satin; front guide bar makes longer underlaps at the back side, while the back guide bar makes the usual single-bar tricot construction. Basic difference between the lapping movements of satin and loop raised fabric is movement of guide bars, which is in opposite direction in case of satin and other of its kind fabrics, while both front and back guide bar fabrics move in the same direction for underlaps for loop raised fabrics. These structures are unstable, and loop does not lie vertical but usually is inclined at certain angle left and right in different courses. The process of loop raising makes the fabric stable and wales lines in vertical direction. Lapping movement of loop raised fabric is shown in Figure 10.31. Numerical notation of front guide is 1-0/3-4// or 1-0/4-5//, and back guide bar has 1-0/1-2//.

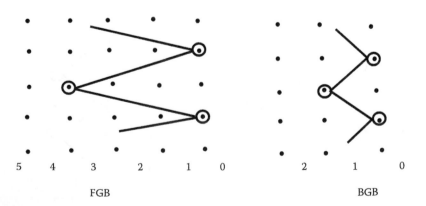

FIGURE 10.31
Lapping movement of loop raised fabric.

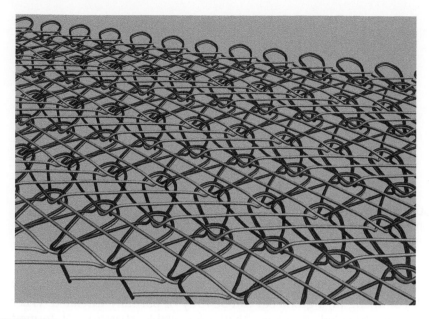

FIGURE 10.32
Technical back side of velour.

10.11.10 Velour or Velvets

Normally, a longer underlap is produced with front guide bar of usually 6–8 needles space. This will produce a float on back side of fabric. This float is brushed in finishing to produce pile. Usually, for strength 40–60 denier nylon or polyester yarn is used on back guide bar, while on front guide bar 55–100 denier acetate or viscose yarn is used, which is converted into pile in finishing by brushing. Shrinkage is in the range between 35% and 50%. In apparel application, popular areal density is 150 g/m². Heavier fabrics made with nylon are usually used for furnishing. The technical back side of velour is shown in Figure 10.32.

10.12 Laying In

Laid-in structures are produced when a guide bar only underlaps and has no overlap (Paling 1952). These structures are produced with yarn that has specific reasons to not knit rather used as an inlay yarn. For laying in, no special or extra equipment is needed; rather, special lapping movement is required. No swinging movement only shogging movement produces laid-in structures. Some of the principles of laying in are as follows:

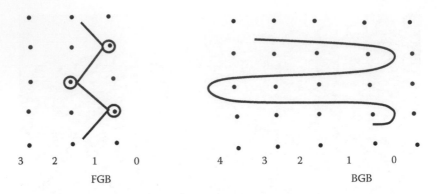

3   2   1   0   4   3   2   1   0

FGB   BGB

FIGURE 10.33
Lapping movement of laid-in structure.

1. Laying is done usually by back bar, and all guide bars can inlay yarn other than front guide bar. Front guide bar produces main structure.
2. For laying in, a guide bar can be fully or partially threaded. When fully threaded, it means that more stable structure is formed. Guide bar can be threaded fully or partially as per fabric design requirement.
3. An inlaid yarn remains only back side of needle, so underlap is produced. These yarns never enter the hook of needle.

Technically, laying-in technique offers some advantages which otherwise are not possible. This can allow yarns that cannot be knitted into loops to be used as an inlaid yarn. Those yarns can be laid in that are coarser and exceed the limits dictated by the machine gauge. Less yarn is used when it is laid in, and thus more expensive patterning yarns can be used in this way. Range of design can be broadened with less yarn limitations when yarn is laid in. Figure 10.33 shows the lapping movement of inlaid structure. Front guide bar has 1-2/1-0// movement, while laying-in thread makes 0-0/4-4// lapping.

10.13 Open Work Effect by Laying In

10.13.1 Tulle

Mesh structures are used according to design requirement, sometimes used alone and sometimes used as a ground for different designs produced. There are three gauges used for raschel mesh: E14, E18, and E24. Different mesh structures are produced depending on design and guide bars movement

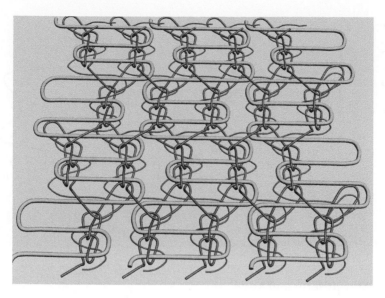

FIGURE 10.34
Five course tulle structure.

including three-course tulle, cross tulle, 3/2 tulle, and five-course tulle. Figure 10.34 shows five-course tulle structure, and Figure 10.35 shows its lapping movement. Movement of front guide bar is 2-0/0-2/2-0/2-4/4-2/2-4//; for inlay ground, it is 0-0/2-2/0-0/4-4/2-2/4-4//; and for inlay pattern, it is 0-0/4-4/0-0/4-4//.

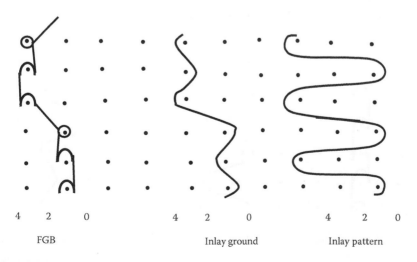

4	2	0		4	2	0		4	2	0
	FGB				Inlay ground				Inlay pattern	

FIGURE 10.35
Lapping movement of five course tulle structure.

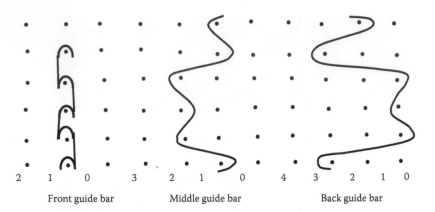

FIGURE 10.36
Lapping movement of Marquisette.

10.13.2 Marquisette and Voile

These structures are named after woven construction. Fully threaded guided bars are used to produce these structures with front guide bar that makes the pillar stitches. These two meshes are different from each other based on the direction of movement of laying-in bar. In Marquisette, inlays form a square mesh, while in voile, a diagonal mesh is formed. Figure 10.36 shows the lapping movement of marquisette, and numerical notation is as follows:

Front guide bar 1-0/0-1//
Middle guide bar 0-0/2-2/1-1/2-2/0-0/1-1//
Back guide bar 3-3/0-0/1-1/0-0/3-3/2-2//

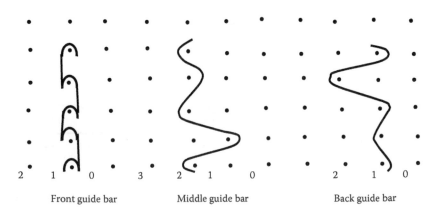

FIGURE 10.37
Lapping movement of voile.

Lapping movement of voile is shown in Figure 10.37, and numerical notation is as follows:

Front guide bar 1-0/0-1//
Middle guide bar 2-2/0-0/2-2/1-1/2-2//
Back Guide bar 0-0/1-1/0-0/2-2/0-0//

10.13.3 Elastane Fabrics/Power Nets

Elasticized fabrics are used for both technical and daily-use applications like corsetry, swimwear, lingerie, and leisure wear. Elasticized fabrics are knitted on both tricot and raschel knitting machines. While using elastane yarn, care is needed as tension variations are very important so need to be controlled. Power net is a popular structure of this class of fabrics. Four guide bars are used for this. Two are used for Nylon yarn feeding, and two are used for elastane yarn. Figure 10.38 shows the lapping movement of power nets, and numerical notation is as follows:

First guide bar 2-4/4-2/4-6/4-2/2-4/2-0//
Second Guide bar 4-2/2-4/2-0/2-4/4-2/4-6//
Third guide bar 2-2/0-0//
Fourth guide bar 0-0/2-2/

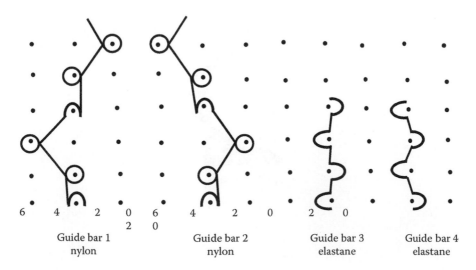

| | Guide bar 1 nylon | Guide bar 2 nylon | Guide bar 3 elastane | Guide bar 4 elastane |

FIGURE 10.38
Lapping movement of power nets.

10.13.4 Co-We-Nit

This term was introduced to describe the knitted structures that have woven-like appearance. So, Co-We-Nit fabrics are technically those fabrics which are knitted in reality but close to woven in appearance. The term Co-We-Nit is used for both fabrics and machines producing these kinds of fabrics. This is a modified form of warp knitting in which pillar stiches are held through by laid-in yarn using fall-plate technique.

10.14 Full-Width Weft Insertion

Weft insertion in warp knitting is done both for technical and esthetic reasons. Weft insertion is like laying in, but unlike laying in, weft is inserted from selvedge to selvedge as weft is inserted in woven fabric to cover full width of fabric. Weft is not allowed to enter the hooks of needles; rather, it remains back side of needle so appear as an underlap. The purposes of weft insertion are to increase fabric lateral stability; to increase the fabric cover; to use fancy yarn for ornamentation purpose; to use such a yarn that cannot be knit due to some technical reasons like glass, carbon, and other yarn;

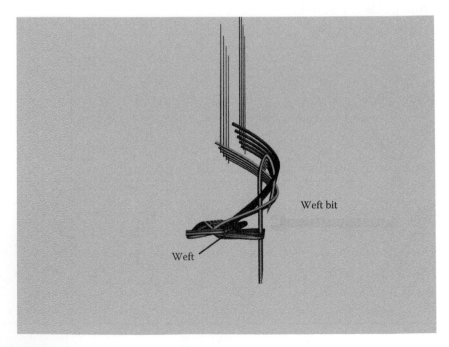

FIGURE 10.39
Weft insertion mechanism on warp knitting machine.

and to impart some functionality in fabric as per end use requirement. Weft is inserted to machine by a special mechanism which inserts yarn to full width of fabric as shown in Figure 10.39. Weft is supplied from a creel placed at back of machine. Both tricot and raschel machines are equipped with weft insertion system these days. When the needles are in the lower position during the warp knitting cycle, an open-shed effect is created at the back of the machine. It is then possible for a weft yarn, laid across the full width of the machine, at that time weft is to be carried forward by special weft insertion bits over the needle heads and deposited on top of the overlaps on the needles and against the yarn passing down to them from the guide bars. In this way, the inserted weft will become trapped between the overlaps and underlaps in the same manner as an inlay yarn when the needles rise. The weft will run horizontally across the complete course of loops.

10.15 Multiaxial Knitting

There are some applications where we need such a fabric/reinforcement that requires force bearing all direction. Traditional fabrics woven and knitted do not fulfill these requirements, but even some advanced structures like full-width weft inserted warp knitted fabric and Co-We-Nit fabrics cannot fill this gap. So, multiaxial warp knitted structures are used to fill this gap.

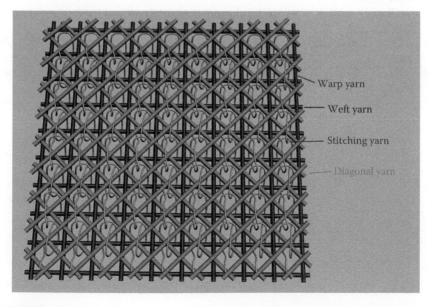

Warp yarn

Weft yarn

Stitching yarn

Diagonal yarn

FIGURE 10.40
Multiaxial warp knitted fabrics.

In multiaxial warp knitting, yarn is used in different directions: main yarn, a weft yarn, and two sets of yarn in bias direction as shown in Figure 10.40.

The pillar stitches are made through the front guide bar, incidentally guide bar number 1 of a raschel machine. There are two sets of inlay yarns, one along the width of the fabric and the other along its length. Inlays along the width are inserted by the magazine weft insertion system, while the ones along the length result out of mislaps (0 0//). The mislapped threads occupy a position closest to the loops knitted by the front bar which are visible on the technical front side of the fabric. The inlays along the two bias directions are accordingly controlled by guide bars 2 and 3. The inlays along the bias directions move through four needles space over four knitted courses. Hence, the extent of lateral shogging is over one needle space for every course. However, the two sets appear to keep moving continuously in opposite directions without reversing after a certain number of courses, as is the case with normal inlays. This means that while Guide Bar 2 keeps moving from left to right, bar 3 moves form right to left.

References

Ajgaonkar, D.B. 1998. *Knitting Technology*. Universal Publishing Corporation, Bombay.

Paling, D.F. 1952. *Warp Knitting Technology*. Harlequin Press Company Limited, Manchester.

Spencer, David J. 2001. Knitting technology. 3rd edn. Woodhead Publishing Limited, Cambridge.

AU, K.F. 2011. *Advances in Knitting Technology*. 1st edn. Woodhead Publishing Limited, Cambridge.

Anbumani, N. 2010. *Knitting Fundamentals, Machines, Structures and Developments*. New Age International (P) Limited, New Delhi, India.

Ray, S. Chandra. 2011. *Fundamentals and Advances in Knitting Technology*. Woodhead Publishing India Pvt. Ltd, New Delhi, India.

11

Color and Stitch Effect

Habib Awais

CONTENTS

Fashion is normally composed of five elements. Color is one of them and the remaining are texture, pattern, style, and silhouette [1]. There are various stages where one can introduce ornamentation for design purposes: fiber, yarn, weaving or knitting, dyeing, and finishing. Color can be introduced at fiber, yarn, and fabric levels. Colored fibers are mixed in blow rooms to produce mélange yarns. Colored yarns or rovings are twisted together to produce marl yarns. Similarly, color can be introduced in fabrics by applying various dyes in the processing. Fancy yarns as well as yarns manufactured by various spinning techniques can be combined and used to produce a special effect in a fabric [2]. The dyeing process provides the possibility of differential and cross-dyeing of fabrics composed of more than one type of fiber or yarn [3]. Finishing and printing processes may utilize chemicals or heat to introduce color design in plain surfaces [4]. Embroidery is also used to produce special designs and motifs in garments [5].

11.1 Techniques Involved to Achieve Different Color Effects

In knitting, stitches are used for creating various effects. But there are limitations in creating effects by stitches due to low machine speed, less feed

utilization, less efficiency, and time-consuming process of design change. Other techniques used to produce design in knitting apart from stitches are plating, intarsia, horizontal and vertical stripes, and jacquard. Each technique has its own strengths and weaknesses.

11.1.1 Plating

Plating is mostly used for interlock fleecy, plush, and single jersey fabrics. Great precision is required to produce plating and limited color can be offered in this technique. Two different color yarns are used in plating techniques. One of them acts as a main color yarn and the other is an alternate color yarn. These two yarns are knitted in a stockinette stitch. The main color yarn is knitted on front having the alternate color yarn behind. The process of knitting together two yarns in a stockinette stitch creates a fabric that is lined. The alternate color of yarn may be seen through from the other side of the main color yarn, which causes a variation in the color of the fabric. Plating with two yarns is often done with a rough-textured main color and a softer alternate yarn to create a lining in the fabric that is not scratchy to the touch. The most common use of plating techniques is in creating motifs during socks manufacturing on electronically controlled, single cylinder knitting machines, as shown in Figure 11.1.

There are two types of plating: *reverse plating* and *sectional plating*. In reverse plating, specially shaped sinkers or yarn feed guides are used for change over positions of contrast color yarns at the needle head by their controlled movement. In sectional plating, the ground yarn knits continuously across the full width while the plating carrier tubes, set lower into the needles, supply yarn in a reciprocating movement to a particular group of needles, so that the color shows on the face [6].

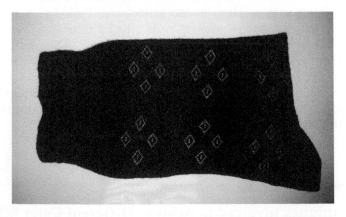

FIGURE 11.1
Plated socks.

Spandex is often added to yarns to increase their shape retention, especially for cotton, ramie, and rayon blends. It can be combined with any fiber. There are various methods of combining the spandex; the easier is through a technique called plating. The standard yarn is knitted to the outside and it fully conceals the spandex yarn on the back because it is larger. The spandex yarn is fed into a machine with a tensioning device, so it is fully stretched. The addition of spandex works well with fine gauges. The plating technique using two yarns at once is employed not just with spandex but for other fibers as well.

11.1.2 Intarsia

Motifs with multiple colors are created in knitted fabrics by using the intarsia technique. Unlike other multicolor techniques (including Fair Isle and double knitting), there is only one "active" color on any given stitch, and yarn is not carried across the back of the work; when a color changes on a given row, the old yarn is left hanging. The intarsia carrier is a separately available attachment for most knitting machines that aids in the knitting of geometric and argyle designs, but also ideal for any multicolor large designs (sometimes called "picture knitting"). The intarsia carrier takes care of the long floats that would appear on the back of the fabric. Several colors can be knitted in one row without floats. Knitting in intarsia theoretically requires no additional skills beyond being generally comfortable with the basic knit and purl stitches. Materials required for manual knitting include multiple colors of yarn, standard needles, and *bobbins*. Bobbins serve to contain the inactive yarn and help keep it from getting tangled. Unlike the narrow, wooden ones used to make bobbin lace, modern intarsia bobbins resemble translucent plastic yo-yos that can snap tight to prevent the yarn from unwinding.

After winding a few yards of each color onto its own bobbin (and possibly several bobbins' worth of some colors), the knitter simply begins knitting its pattern. When they arrive at a point where the color changes, the knitter brings the new color up underneath the old one (to prevent holes) and starts knitting with it. If flat knitting, at the end of the row, the piece is turned round just as with regular knitting, and the knitter returns the way they came. The simplest intarsia pattern is for straight vertical stripes. After the first row, the pattern is continued by always working each stitch in the same color as the previous row, changing colors at exactly the same point in each row. To make more elaborate patterns, one can let this color boundary drift from row to row, changing colors a few stitches earlier or later each time. Intarsia patterns are almost always given as charts (which, because of the mechanics of knitting, are read from beginning of lower right and continuing upward). The charts generally look like highly pixelated cartoon drawings, in this sense resembling dot-matrix graphics or needle point patterns (though usually without the color nuance of the latter).

11.1.3 Jacquard Design

Jacquard technique is commonly known as individual stitch selection [6]. This is the most widely used method for producing design in knitted fabrics. Individual stitch is selected as per choice or design. Latch needles are the most suitable for this technique due to their self-acting nature. Two-color and three-color jacquards are mostly used to introduce special effects. Motif, pictures, and geometric shapes are easily introduced in fabrics using jacquards. But there is a limitation of float in single jersey fabrics, which can be overcome by using the accordion technique. In the accordion fabric, long floats are held in by tuck stitches. For this purpose, needles required an extra butt and a tuck cam after the pattern wheel selection. Tuck stitches are introduced on all odd needles, which are not being knit, odd and even needles combination on nonknit needles, or selective selection of nonknit needles.

11.2 Horizontal and Vertical Stripes

There are two methods for producing horizontal stripes in knitting: auto stripes and feed stripes. In auto stripes, a machine is equipped with special equipment or attachment for the selection of required yarn from the number of available colored yarns. In feed stripes, colored yarns are carefully placed on feeders in such a way that stripes are produced as shown in Figure 11.2.

Vertical stripes are also a source of creating special effect in knitted fabrics. Vertical stripes are produced normally by the special arrangement of yarn on feeders as well as cam arrangements. Knit stitch is used for the yarn whose stripe is required to be produced. By miss stitch, the other color will hide on the back of the fabric. Only that color will come on front for which

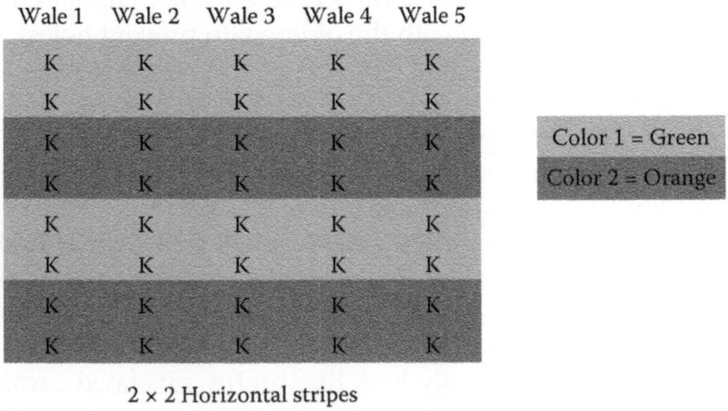

Wale 1	Wale 2	Wale 3	Wale 4	Wale 5
K	K	K	K	K
K	K	K	K	K
K	K	K	K	K
K	K	K	K	K
K	K	K	K	K
K	K	K	K	K
K	K	K	K	K
K	K	K	K	K

Color 1 = Green
Color 2 = Orange

2 × 2 Horizontal stripes

FIGURE 11.2
Horizontal stripes.

Wale 1	Wale 2	Wale 3	Wale 4	Wale 5	Wale 6	Wale 7	Wale 8
K	K	M	M	K	K	M	M
M	M	K	K	M	M	K	K
K	K	M	M	K	K	M	M
M	M	K	K	M	M	K	K
K	K	M	M	K	K	M	M
M	M	K	K	M	M	K	K

Color 1 = Yellow
Color 2 = Red

2 × 2 vertical stripes

FIGURE 11.3
Vertical stripes.

the knit stitch is used as shown in Figure 11.3. Horizontal and vertical stripes are also produced in backing for rib jacquard designs. In these designs, the face of the fabric contains a jacquard design, and on the back of the fabric, stripes of specific width are produced by using electronic control.

11.3 Twill Effect

The twill effect basically produced diagonal lines in the fabric. The twill effect is produced by the combination of knit and miss stitches on even and odd needles but this odd, even behavior changes on their respective colored yarn, as shown in Figure 11.4. First, all even needles of Color A produce knit

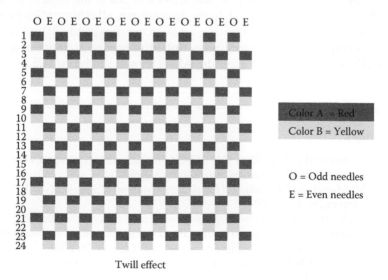

Color A = Red
Color B = Yellow

O = Odd needles
E = Even needles

Twill effect

FIGURE 11.4
Twill effect.

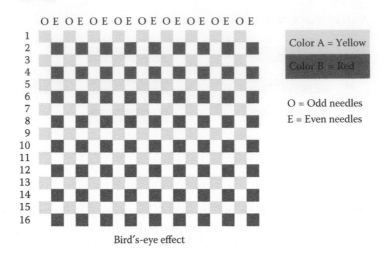

Bird's-eye effect

FIGURE 11.5
Bird's eye effect.

stitches for the first course, and for the second course, all even needles produce knit stitches for Color B [7].

11.4 Bird's Eye

Bird's eye is a combination of knit and miss stitches at alternative position, which shows the bird's-eye effect. In the first course, all odd needles produce knit stitches, and in the second course, all even needles produce knit stitches. This odd, even combination continues until the repeat is completed [6]. In this way, the bird's-eye effect is produced, as shown in Figure 11.5.

References

1. B. J. Collier and J. R. Collier. CAD/CAM in the textile and apparel industry. *Clothing and Textile Research Journal*, 8(3), 7–13, 1990.
2. E. S. Meadwell, *An Exploration of Fancy Yarn Creation*, MS thesis, North Carolina State University, USA, 2004.
3. Y. Nawab (ed.). *Textile Engineering: An Introduction*. De Gruyter, Berlin, 2016.
4. P. R. Wadje, Textile–Fibre to Fabric Processing. *Journal of the Institution of Engineers (India), Part TX: Textile Engineering Division*, 90:28, 2009.
5. E. R. Post, M. Orth, P. R. Russo, and N. Gershenfeld. E-broidery: Design and fabrication of textile-based computing. *IBM Systems Journal*, 39(3 and 4), 840–60, 2000.
6. D. J. Spencer, *Knitting Technology*, CRC Press, New York, 2001.
7. C. Iyer, B. Mammel, and W. Schach, *Circular Knitting*, Bamberg, Germany, 2004.

Index

Printed and bound by CPI Group (UK) Ltd, Croydon, CR0 4YY

24/10/2024

01778301-0009